Computing
for Biologists

An introduction to BASIC programming
with applications in the life sciences

Computing for Biologists

An introduction to BASIC programming with applications in the life sciences

ALAN FIELDING

THE BENJAMIN/CUMMINGS PUBLISHING COMPANY INC.

Menlo Park, California · Reading, Massachusetts · Wokingham, England
Don Mills, Ontario · Amsterdam · Sydney · Singapore · Tokyo
Mexico City · Bogota · Santiago · San Juan

© 1985 Addison-Wesley Publishers Limited
© 1985 Addison-Wesley Publishing Company, Inc.

The cover design shows graphical representations of the protein insulin, based on the refined 2-zinc insulin models of Isaacs and Agarwal (Isaacs, N.W. and Agarwal, R.C., *Acta Crystallogr. A*, 34 (1978) 782). Computer graphics produced by the Graphics Systems Research Group, IBM UK Scientific Centre.

Photoset direct from the author's text by Quorum Technical Services Ltd.
Printed in Finland by Werner Söderström Osakeyhitö. Member of Finnprint.

Library of Congress Cataloging in Publication Data
Fielding, Alan
  Computing for biologists.

  Bibliography: p.
  Includes index.
  1. Biology – Data processing. 2. Basic
(Computer program language) I. Title.
QH324.2.F54 1985   001.64'24   84-24525
ISBN 0-8053-2515-8

ABCDEF 8987654

Contents

Foreword   ix

Chapter 1   Introduction to Biological Computing   1

    1.1   Biological applications of computers   1
    1.2   Elements of computer science   3
    1.3   Number systems   4
    1.4   Computer hardware   5
    1.5   Summary   12

Chapter 2   Programming Languages   13

    2.1   Introduction   13
    2.2   Machine language programs   14
    2.3   High level languages   16

Chapter 3   Introduction to BASIC   20

    3.1   Program development   20
    3.2   BASIC programs   21
    3.3   The problem   22
    3.4   The program structure   23
    3.5   The complete program   31
    3.6   Summary   32
    3.7   Problems   33

Chapter 4   Loops and Arrays   35

    4.1   The simple FOR ... NEXT loop   35
    4.2   Numeric arrays   40
    4.3   Nested FOR ... NEXT loops and two dimensional arrays   45
    4.4   Summary   50
    4.5   Problems   51

Chapter 5   Functions   52

    5.1   Outline of BASIC functions   52
    5.2   The use of BASIC functions in programs   53
    5.3   Example programs using BASIC functions   56
    5.4   User defined functions   60
    5.5   Summary   61
    5.6   Problems   62

Chapter 6	**Program Control Structures**	**64**
6.1	Introduction	64
6.2	Unconditional commands	64
6.3	Conditional commands	67
6.4	Summary	75
6.5	Problems	76

Chapter 7	**Strings**	**77**
7.1	Introduction	77
7.2	BASIC string commands	77
7.3	Print formatting	82
7.4	Example programs	83
7.5	Summary of commands	86
7.6	Problems	87

Chapter 8	**Disks, Files and Operating Systems**	**89**
8.1	Introduction	89
8.2	Disks	89
8.3	Files	93
8.4	Operating systems	97

Chapter 9	**Computer Graphics and Image Processing**	**102**
9.1	Introduction	102
9.2	Hardware review	102
9.3	Programming principles	106
9.4	Biological applications	110
9.5	Summary	114

Chapter 10	**Structured Programming**	**115**
10.1	Historical background	115
10.2	What is structured programming?	115
10.3	Structured commands in BASIC	116
10.4	Writing structured programs	117
10.5	Example	119
10.6	Summary	123
10.7	Problems	123

Chapter 11	**Computer Models**	**124**
11.1	Introduction	124
11.2	Model 1 – energy balance in living organisms	124
11.3	Model 2 – selection against a recessive allele	131
11.4	Model 3 – competition between two species of animal	134
11.5	Summary	139

Chapter 12	**Information Technology**	**140**
12.1	Introduction	140
12.2	Computer communications	140
12.3	Local area networks	142
12.4	Databases	144

12.5	Online information retrieval	146
12.6	DNA databases	149
12.7	Summary	151

Chapter 13	**Working with Mainframe Computers**	**152**
13.1	Introduction	152
13.2	MINITAB	154
13.3	SPSS-X	156

Chapter 14	**Program Optimisation**	**161**
14.1	Introduction	161
14.2	Optimisation of BASIC programs	162

Chapter 15	**Summary**	**165**

Appendix A	**Number Storage Systems and Sources of Error**	**169**

Appendix B	**ASCII Conversion Table**	**174**

Appendix C	**Microcomputer Interfacing**	**176**

Appendix D	**Answers to selected problems**	**180**

Appendix E	**Bibliography**	**185**

Index		187

Foreword

My initial aim, in writing this book, was to provide an introduction to computer science, with particular emphasis on those areas which should be of greatest value to the majority of biologists. There is little doubt that computer literacy is going to become an important skill for practising and student biologists from all areas of the life sciences. This book was, however, never intended to be a comprehensive treatise on the uses of computers in the biological sciences. I hope that my book can be used to overcome the initial inertia which many biologists express towards computers. In many respects this inertia is similar to the enmity which is often directed at biomathematics by biology students. Frequently these difficulties arise because biologists are taught these 'peripheral' subjects by staff who do not have a biological background. Since I am primarily a biologist I hope that I can offer a more sympathetic approach to the subject of biological computing.

The book is not based on any specific model of computer. Most of the example programs, which are written in the BASIC programming language, should work on the majority of computers without amendments. This approach was adopted because there are already a large number of books dedicated to specific computers. Also, many readers will have access to large mainframe computers. These larger computers often use versions of the BASIC language which are quite different from those found on microcomputers. These mainframe dialects frequently place constraints on the programmer which do not exist in microcomputer versions of BASIC. It would be an impossible task to test the programs on all of the popular computers. Consequently, if you have a problem with any of the programs, consult your manual and examine the syntax of the commands which have been used.

Computer science is not a subject which can be learned entirely from books. The only reliable method of developing an understanding of computers is to use one. In computing jargon (of which there is lots) this is known as gaining 'hands-on' experience. Do not be afraid of making mistakes. Unless you are delving around inside the computer you will not damage it. Making and correcting mistakes is one of the best methods available for learning about computers. Consequently, I would recommend that whenever it is possible you should type in and run the example programs. You will probably make some typing errors as you do so. These typing errors will prevent the program from working correctly, but identifying and correcting the errors will be a valuable experience. It is an unfortunate fact that, as the example programs become more complex, the level of biological and/or mathematical knowledge required also increases. It is not too early to learn that computers cannot be used as a substitute for biological knowledge.

I would like to thank the many people who have helped me to write this book. In particular I would like to express my gratitude to the following colleagues in the Department of Biological Sciences at Manchester Polytechnic: Dr Chris Smith, who has read and constructively criticised everything that I have written for this book; Dr Derek Gordon for much helpful advice and useful discussions; Dr Stan Shaw who explained, and allowed me to adapt, his energy balance practical; Dr Ian Graham for his report on the DES conference on Microcomputers in Biology in Higher Education. I am also grateful to Derek Scoular for his useful criticisms. I would also like to thank Chiltern Electronics of Chalfont St Giles who were very helpful on several occasions, but particularly when my computer failed at a very crucial stage during preparation of the manuscript.

I would like to express my gratitude to the three reviewers: Dr J. D. Spain, Dr A. N. Barrett and Dr C. G. Moore, who offered many helpful suggestions and criticisms; and to Dr Jane Burridge of IBM who supplied the cover photographs. I am very grateful to the staff at Addison-Wesley for their enthusiasm and help over the past twelve months.

Finally, I would like to thank my family, particularly my wife Sue and brother David, without whose help and patience I could not have completed the work.

Alan Fielding
Bolton, October 1984

Chapter 1 Introduction to Biological Computing

A question asked by many biology students is 'why should I study computing?'. The answer is obvious when you consider that most advances in biological science have depended upon the development and application of new technologies and equipment. A good example of this is provided by the advances in our understanding of the internal structure of cells that were made following the introduction of the transmission electron microscope. It is quite likely that recent developments in computers will provide the opportunities for further advances in many areas of biology. If this does occur then computers may become as commonplace in biological laboratories as the light microscope! As with most of the complex equipment used by biologists only an understanding of the basic principles of operation is necessary. It is more important that the biologist is aware of the computer's potential as a tool in the investigation of biological problems. This book will introduce some elements of computing which should be of value to biologists, particularly undergraduates, who are new to computing and so enable them to make use of this important new technology and develop the level of computer literacy that is going to become an important skill for biologists in the future.

1.1 Biological applications of computers

Although existing biological applications of computers are extremely diverse, there are still many areas in which they could be applied most successfully. The following list is not intended to be a comprehensive survey of the current uses but aims to illustrate the main areas in which computers are of value to biologists.

Data analysis
Small microcomputers can be used for the rapid analysis of routine data sets in the laboratory, for example, in the calculation of the Michaelis constant in biochemistry. Computers also facilitate the rapid and detailed analyses of large data sets such as those produced in many areas of ecological and behavioural work. This can be important because the rapid assessment may suggest new hypotheses which may not have been identified previously since the time required for such analyses would have been prohibitive.

Teaching
The use of computers in the teaching of biology falls into two main categories. The first, and probably most common, involves the development

of simple computer models of complex biological systems (for example, species interactions, enzyme reaction mechanics, population genetics). These models are useful because they allow a student to manipulate the important factors controlling the system (for example, competition coefficients, substrate concentration, migration rates) and they will then predict how changes would affect the system. In this way data is made available for analysis and discussion that would not normally be available because of the equipment and time required. The second area is one which may grow in importance as computers become more commonplace and is concerned with Computer Aided Instruction (CAI). In CAI computers are used as teaching machines. A student is presented with information and then asked a series of increasingly more difficult questions. If an incorrect reply is made the computer should provide additional information and explain why the reply was incorrect. The more advanced CAI packages also include facilities which allow the progress of a student to be monitored; in this way the teaching can be adapted to the needs of individual students.

Databases
A database is a collection of information that is stored in a form accessible by computers. For example, the large and rapidly expanding body of information about DNA base sequences can only be exploited successfully when the information is stored in a computer database. Databases are therefore potentially very flexible and it is for this reason that their uses will become more diverse and commonplace. A database can be established on several levels, ranging from the personal to the international, depending upon its intended use. The concept and uses of databases will be discussed in Chapter 12.

Analysis of complex structures
Biologists often deal with materials which have very complex structures (for example, macromolecules, cell organelles, embryonic development patterns). An understanding of the factors which control these structures is exceedingly difficult without the aid of computers. Computers are useful for such studies because of the vast amount of information which they can store and because of their speed in analysing it.

Computer models
In common with all scientists, biologists often find it necessary to produce a simplified version or model of the system under investigation. Computers are ideal tools for the development of such models because changes in various parameters which affect the model can be assessed rapidly. Computer models can, therefore, be used as a mechanism to evaluate hypotheses which can then be tested by subsequent experimentation. It is, however, very important to remember that the accuracy of any model depends, to a large extent, upon an understanding of the underlying principles. Good biological models cannot be developed if the biological knowledge is lacking.

Word processing
Word processing involves the use of computers to handle text rather than numbers. The ability of computers to manipulate text results from the

development of computer programs dedicated to that task. Word processors can be used to prepare manuscripts and reports. The advantage of word processors over normal typewriters is that they speed up the revision and correction procedures and therefore reduce the amount of time required for such activities.

Equipment control
This is a very important area which will continue to grow. Computers can provide the facilities for the automatic control of equipment so that a required sequence of procedures will be followed including data collection and analysis. In theory, most electronic laboratory equipment can be connected to a computer but there are many problems that must be overcome. These will be introduced in Appendix C.

Some of the most recent uses of computers in the biological sciences are illustrated in a book edited by Geisow and Barrett (see Appendix E for details).

One of the problems which all computer novices must face is the vast array of jargon associated with the subject, and because of the way in which it evolved normal linguistic skills are not always of value in interpreting the meaning of a particular piece of terminology. Computer terminology is frequently derived from acronyms (words formed from the initial letters of other words). Many of these acronyms seem to be excessively contrived. Origins of other words depended upon the characteristics and facilities offered by early computers, many of which are now obsolete or rare, and so the definitions of these original words have changed to take account of new developments. A consequence of this can be confusion over the exact definition of terms. This can make life difficult for you!

1.2 Elements of computer science

There are two main branches of computer science. The first is concerned with computer hardware and the second with software. Hardware means the mechanical and electronic devices from which computers are constructed, while software consists of the sets of instructions or programs which are used to direct the operations of the computer hardware. At present the advancement of biological computing is being restricted more by deficiencies in existing software rather than limitations in current hardware. While many biologists should be capable of developing the skills required to adapt existing hardware it is very unlikely that they will become directly involved in the development of new computer hardware. It is therefore in the development and application of new and existing software that biologists can make their biggest contributions to biological computing. Consequently, this book will concentrate on the software, or programming, aspects of biological computing.

Even though most biologists will be concerned primarily with programming it is still necessary to understand something about the nature of computer hardware and this will be introduced in the following paragraphs. However, before the hardware can be explained, an understanding of how a computer stores information is required.

1.3 Number systems

The primary unit of information in electronic computers is the bit or Binary digIT. A bit can have only one of two states 1 or 0. It is therefore ideal for electronic manipulation since the two states can be represented by two voltage states. A single bit can convey only two pieces of information and it is, therefore, not very useful by itself. Consequently, bits are organised into larger functional groups with which the computer can operate. These groups of bits are known as words and the number of bits in a word is fixed for any particular computer. The most common grouping in microcomputers is 8 bits, which together form a unit known as a byte. The 8 bits in a byte can be arranged into 256 different combinations of ones and zeros. This represents a considerable amount of information. Word lengths of greater than 8 bits are, however, becoming more common on recently developed microcomputers. Most large computers operate on word lengths of 16 bits or longer and in the largest computers this may extend to 60 bits or more. The length of the word is exceedingly important in determining the capability of a computer. In general the most powerful and efficient computers are those which use the longest words. It is quite likely that the standard word length of microcomputers will increase from 8 to 16 or possibly 32 bits in the very near future. This will bring about a vast increase in the power of such machines.

Since the basic unit of computers is the binary digit they are unable to operate directly on the more familiar decimal or denary number system but utilise instead a binary number system. Although not essential, it is advisable that the computer user has some understanding of the various number systems associated with computers.

Each of the 256 different combinations of ones and zeros in a single byte can be used to represent a decimal number between 0 and 255. This relationship can be explained if each bit in the byte is taken to represent a power of two. The eight bits can be numbered, reading from right to left, as 1 to 8 or more usually as 0 to 7. The first bit, bit zero, is equivalent to 2^0, which is one in the decimal system. The second bit, bit one, is equivalent to 2^1, which is two. The third bit, bit two, is equivalent to 2^2 or four, etc. The bit numbering system, 0 to 7, reflects, therefore, the power to which two is raised.

Bit number	7	6	5	4	3	2	1	0
Power of 2	7	6	5	4	3	2	1	0
Decimal equivalent	128	64	32	16	8	4	2	1

Bits may be either on or off (1 or 0), as determined by the voltage state, and therefore it is possible to get different bit patterns made up of zeros and ones. If a bit is off (zero), then that power of two is not included in the calculation of the decimal equivalent. Decimal one would, therefore, be represented by a byte composed of the following bits 00000001, decimal two by 00000010, decimal three by 00000011 etc. The largest decimal number that can be represented by one byte is 255 or 11111111. Combinations of bytes can be used to represent numbers larger than 255. The binary system, although ideal for computers, is a difficult system for direct human manipulation. Consequently, number systems which are intermediate between the binary and decimal systems have been devised.

Humans normally operate a number system which is based on ten: the decimal system. The origins of the decimal system presumably result from our pentadactyl limb, i.e. we have ten fingers. If we had evolved with a tridactyl limb then the normal number system would probably have been based on six. The most common computer word length is eight bits so a number system based on eight, the *octal* system, was developed. However, it was soon realised that a system based on sixteen, the *hexadecimal* system, would be of greater value. One problem that was encountered during its development was how the numbers between decimal 10 and 15 should be represented. The normal decimal numbers (0–9) are used for the first ten hexadecimal numbers, while the letters A–F are used for the remaining five. Table 1.1 shows the relationships between some decimal numbers and their binary, octal and hexadecimal equivalents. The relationship between binary and hexadecimal numbers can be simplified by considering the 8-bit byte to be composed of two 4-bit nibbles! Thus, the byte 01100100 (100 decimal, 64 hexadecimal) is composed of the two nibbles 0110 and 0100. The first nibble is equivalent to the hexadecimal (and decimal) digit 6, while the second nibble is equivalent to the hexadecimal (and decimal) digit 4.

Table 1.1 Relationships between number systems

Decimal	Derivation	Hexadecimal	Derivation	Binary
1	$10 \times 0 + 1$	1	$16 \times 0 + 1$	00000001
2	$10 \times 0 + 2$	2	$16 \times 0 + 2$	00000010
3	$10 \times 0 + 3$	3	$16 \times 0 + 3$	00000011
4	$10 \times 0 + 4$	4	$16 \times 0 + 4$	00000100
5	$10 \times 0 + 5$	5	$16 \times 0 + 5$	00000101
6	$10 \times 0 + 6$	6	$16 \times 0 + 6$	00000110
7	$10 \times 0 + 7$	7	$16 \times 0 + 7$	00000111
8	$10 \times 0 + 8$	8	$16 \times 0 + 8$	00001000
9	$10 \times 0 + 9$	9	$16 \times 0 + 9$	00001001
10	$10 \times 1 + 0$	A	$16 \times 0 + A$	00001010
11	$10 \times 1 + 1$	B	$16 \times 0 + B$	00001011
12	$10 \times 1 + 2$	C	$16 \times 0 + C$	00001100
13	$10 \times 1 + 3$	D	$16 \times 0 + D$	00001101
14	$10 \times 1 + 4$	E	$16 \times 0 + E$	00001110
15	$10 \times 1 + 5$	F	$16 \times 0 + F$	00001111
16	$10 \times 1 + 6$	10	$16 \times 1 + 0$	00010000
50	$10 \times 5 + 0$	32	$16 \times 3 + 2$	00110010
100	$10 \times 10 + 0$	64	$16 \times 6 + 4$	01100100
140	$10 \times 14 + 0$	8C	$16 \times 8 + C$	10001100
255	$10 \times 25 + 5$	FF	$16 \times F + F$	11111111

More advanced aspects of numerical storage systems will be introduced in Appendix A.

1.4 Computer hardware

1.4.1 Introduction

Since the invention of the first true electronic computer in about 1945 there has been a period of rapid expansion and development which has resulted in the wide diversity of present day computers. Three main categories of digital computers are recognised: mainframe computers, minicomputers and micro-computers. Mainframe computers are large and very powerful machines which are capable of supporting many users simultaneously. Although prices have effectively fallen in the last ten years they are still very expensive machines, both to purchase and to maintain and they tend to be found only in large institutions. The boundaries between the three types become blurred particularly with respect to the intermediate sized machines known as minicomputers. The problem has been amplified by the fact that recent developments have produced minicomputers which are considerably cheaper and more powerful than some of their mainframe ancestors. The final category, microcomputers, probably represents the widest range of machines, from small business computers to cheap personal computers. In the very recent past microcomputers could be characterised by the fact that they were capable of supporting only one user and program at a time. However, recent developments in both hardware and software have pro-duced microcomputers which can support several users and programs simultaneously.

Despite this diversity of computers there are still a number of char-acteristics which are common to all machines. All computers are composed of several sub-units which always include the following:

Central Processing Unit (CPU)
input device
storage device
output device.

The relationships between the above sub-units and the information which they operate on is summarised diagrammatically in Fig. 1.1. This diagram also illustrates the primary function of all computers which is information processing. Information is fed into the computer via input devices, the computer processes this information and eventually releases it in a pro-cessed state via the output device.

1.4.2 The CPU

The CPU is that part of the computer which actually carries out the computation. It also controls and coordinates the operations of the com-puter. In most microcomputers the CPU consists of one of a relatively small

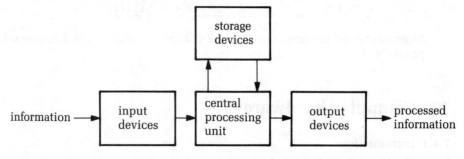

Fig. 1.1 Diagrammatic representation of a computer as an information processing system

range of microprocessors (Table 1.2). Larger computers often have more than one microprocessor and are therefore known as multiprocessor machines. A microprocessor is built around a silicon chip which carries a complex network of electronic circuitry enclosed within a very small volume. The CPU microprocessor can itself be split into a number of distinct functional blocks:

Arithmetic Logic Unit (ALU)
Control Unit (CU)
registers
clock
data and address buses.

The ALU contains circuits which enable it to perform a limited number of relatively simple arithmetic and logical operations which together form the basis for all computing activities. The operations of the microprocessor and ultimately, the computer, require careful control which is maintained by a part of the CPU known as the Control Unit. The CPU may be thought of as a complex switching device which enables operations and data to be routed to the relevant components of the computer.

The registers are special short term memory locations used for the storage of data items and other special information which may be required by the ALU.

Table 1.2 Microprocessors used in the most common microcomputers

Microcomputer	Microprocessor	Data bus	Address bus	Word length
Sinclair Spectrum	Zilog Z80A	8	16	8
BBC Microcomputer	MOS Tech 6502	8	16	8
Commodore Vic 20	MOS Tech 6502	8	16	8
Apple IIe	MOS Tech 6502	8	16	8
Tandy TRS80 Models I-III	Zilog Z80	8	16	8
IBM PC	Intel 8088	8	20	16
ACT Sirius	Intel 8088	8	20	16
Olivetti M-20	Zilog Z-8000	16	23	16
Hewlett Packard Model 16	Motorola 68000	16	24	16

Two solutions have been found to the problem of moving data around within computer systems. The first of these solutions involves the use of what is effectively a single wire. If data in the form of bytes or words is to be transmitted down a single wire the individual bits must be sent sequentially. This in turn means that the computer must have a mechanism by which it can recognise the beginning and end of each byte or word. In most 8-bit computers this is achieved by transmitting one start bit, followed by the eight data bits, and finally two stop bits. Therefore, each byte will be transmitted as a group of 11 bits. This type of data movement is known as serial transmission and is ideal for moving data over long distances. It has the disadvantage of being a relatively slow process. The second solution, to the problem of data movement, uses parallel as opposed to serial bit transmission. A flat cable consisting of as many individual wires as the word length is used to transmit data in parallel. Each bit in the word is carried on a separate wire. In this way the entire byte or word can be transmitted simultaneously. Parallel transmission allows for very rapid data movements but over long distances synchronisation between the bits begins to break down and they may arrive at different times. Consequently, parallel data movements are only used over short distances. The microprocessor uses special parallel data carriers which are known as buses. Data buses are used to transport data while address buses are used to move details of memory locations.

The bit width of the data bus and the internal registers will have an effect on both the speed and accuracy of computation. In general those CPUs with the widest data buses and internal registers will be the fastest and most accurate. Many of the 16-bit microprocessors currently on the market are so called because of the width of the internal registers used in computation. However, since some have data buses only 8 bits wide (Table 1.2) the CPU must carry out two operations to fill and empty each register, which results in an increase in the time taken for each computation. The Motorola 68000 series microprocessor is often described as a 16-bit micro-processor (see Table 1.2). However, since the internal registers are 32 bits wide, 68000 based machines could be described as 32-bit machines. The 68000 microprocessor is available in a number of versions which differ, mainly, in the width of the data bus. This width is important for two reasons: data buses wider than 8 bits result in a very significant increase in production costs but they also increase the power of the system. The most powerful member of the 68000 family has a full 32-bit data bus and is, therefore, a true 32-bit microprocessor. Since this is equivalent to the type of central processor found in most minicomputers, it is almost certain that the next generation of microcomputers will be much more powerful than those currently available.

An important facility of the CPU is that it is able to locate and use data which is stored in the many, separate memory locations of the internal storage device. This is possible since each memory location is labelled with an unique number or address. Memory addresses start at zero and continue up to the highest address which can be accessed by the computer. The amount of addressable memory depends upon the word length of a computer and the width of the address bus. In most 8-bit microcomputers the address bus is 16 bits wide. This restricts addressable memory to a range between 0000000000000000 and 1111111111111111 binary or 0 to 65535 decimal. This represents a potential storage of 65535 bytes. Computers with a word length greater than 8 bits are capable of addressing much larger blocks of memory depending upon the width of the address bus provided by the CPU. The Motorola 68000 has the widest address bus of the currently available microprocessors. Its 24-bit address bus means that it is capable of addressing an enormous 16 mbytes of memory (1Mbyte = one million bytes or one megabyte). A reduction in the address bus to 23 bits, as in the Z8000 microprocessor, reduces the addressable memory to 8 Mbytes. In computing terminology a block of memory made up of 2^{10} memory locations (1024 decimal) is referred to as 1K of memory. Consequently, a computer with 65535 memory locations is said to have 64K bytes of memory (64 × 1024 bytes).

In all programming languages the program instructions are stored in memory as a sequential list of numbers. Therefore, when execution of the program begins, the first instruction must be brought into the CPU and stored in a special memory location known as the instruction register. Once an instruction is in the instruction register it is decoded by the control unit which initiates the sequence of internal (to the CPU) and external signals which are necessary for execution of the instruction. This sequence of signals can control operations such as the movement of data from memory locations into other CPU registers where the ALU can operate on it. The clock, which may be physically separate from the CPU, is very important in synchronising the sequencing of these signals and data movements. The clock speed (number of pulses per second) determines how fast the operations are completed. One of the major developments in the last few years has been an increase in the clock speed associated with many of the common 8-bit microprocessors, such as the Z80. This has resulted in the appearance of 8-bit computers which can perform many tasks faster than some 16-bit models.

1.4.3 Input devices

All computers have methods for getting information, data and instructions into the system and then converting it into a format, groups of bits, that the CPU can operate on. The devices which collect and convert information are known as input devices.

The most commonly used input device is a keyboard which resembles the standard QWERTY typewriter keyboard. However, in addition to the normal letter and number keys there are always a number of extra keys known as control keys. These control keys issue instructions to the computer and do not produce text. One of the most important control keys is one which is usually labelled RETURN or ENTER. As material is typed at the keyboard it is not sent directly to the main part of the computer. It is held in a small area of memory known as a buffer. Only when the RETURN or ENTER key is pressed is the information in the buffer transferred to the main part of the computer.

As computer keyboards become more commonplace the need to acquire adequate keyboard skills will also increase. Much of the speed of a computer is nullified when the operator has to search the keyboard for the position of the next letter or number.

Information typed at the keyboard is converted into a format which the CPU can use. This is achieved by a variety of microprocessors that detect which key has been pressed and then produce a unique sequence of bits corresponding to that character. The sequence of bits used depends upon which of the two main coding systems is being used. Unfortunately these two main coding systems are not compatible. This can create problems when computers using different code systems attempt to communicate with each other. The first of these codes, Extended Binary Coded Decimal Interchange Code (EBCDIC), was developed by IBM for use on their computers. However, the most commonly used system on microcomputers is the ASCII system. This is an acronym for American Standard Code for Information Interchange and was developed by the American National Standards Institute (ANSI). The standard ASCII system is a 7-bit code which uses bit numbers 0–6. The eighth bit (bit 7) is used for a variety of purposes. In this 7-bit code the 128 possible binary numbers (or their decimal equivalents 0–127) are used to represent letters, numbers and a variety of control statements, such as carriage return and line feed. The decimal numbers 128–255 are only available if an 8-bit code is used. Unfortunately most computer manufacturers use the eighth bit in different ways, frequently for the representation of various graphics shapes. The complete 7-bit ASCII code is given in Appendix B.

There are many other methods used for direct input of information into computer systems. Many of these are encountered in everyday life, for example, bar code readers. All of these input devices use a system which is basically the same as that used with keyboards. The information is converted from its original form such as a bar code or characters printed in magnetic ink (often used on bank cheques) into a sequence of electronic signals which the computer can use. This conversion is achieved by using electronic interfaces which act as a link between the 'real' world and the computer. These interfaces always include special microprocessors known as Peripheral Interface Adaptors (PIAs) which carry out the conversion.

It is often necessary or convenient to hold data in a temporary format before entering it into the computer system. The information is held by backing store devices which make use of a variety of media. Temporary storage of information was originally in the form of holes punched into rolls of paper tape or single cards. This information can be transferred to the computer via paper tape or card reading equipment which use a variety of electrical and mechanical devices to extract the information contained in the sequence and pattern of holes. This type of input system is restricted mainly to larger computer systems. More modern methods use magnetic media, in the form of either disks or tape, for data storage. Again specialised equipment is required to convert the signals stored on the magnetic medium into the electronic signals used by the computer, in the same way that audio tape recorders convert magnetic information into music. The main features of the magnetic backing store devices are described below.

1.4.4 Storage devices

An important characteristic of all computers is that they are able to store information in special areas known as memories. The memory which is available to a computer can be split into two categories, main or internal memory and backing store. The internal memory is used by the CPU directly. In most modern systems it is composed of silicon chips, similar in design to the CPU. Most computer systems contain two types of internal memory which are identified by the rather misleading acronyms ROM and RAM. ROM or Read Only Memory is used to store information which must be present in the system at all times. It is therefore used to store programs such as the operating system and BASIC language interpreter (see Chapter 2). The information held in ROM is retained when the computer is switched off and it cannot be easily erased or altered, hence information can only be read from this memory. Reading information from memory is exactly the same as reading words from a page, in that the act of reading does not remove the words from the page. RAM or Random Access Memory is used to store information and programs which are generated by the user. A more correct description of this memory would be Read and Write memory since information can be both read from and written into RAM. Unfortunately Read and Write memory does not provide a convenient acronym! RAM

10

memory is volatile in that when the computer is switched off all of the information contained in RAM is lost. Because of the design used in the production of memory chips any of the individual memory locations is equally accessible by the CPU, i.e. access to the memory is random and not sequential. The distinction between ROM and RAM that is implied by their names is not valid since both ROM and RAM are random access memories.

Backing store or external memory is used to provide more permanent storage for large volumes of data and programs. At the present moment most of this backing store is based upon magnetic media which use either disks or tape, although there is a possibility of devices which use either magnetic 'bubbles' or lasers in the not too distant future.

Magnetic disks come in a variety of sizes but are all based on roughly equivalent technologies. The disks are covered in a magnetic medium which is organised into a number of concentric tracks which produces a superficial resemblance to gramophone records. Information can be either read from or recorded (written) onto the disks by read/write heads similar to those found in domestic tape recorders. The original disks were quite large and made of rigid material, hence the name, hard disks. Hard disks can store very large amounts of information and if implemented on a microcomputer system the available storage space may appear to be infinite! Large computers frequently use disk stores composed of several hard disks, mounted on a central spindle, which can store enormous quantities of information. Hard disks are very expensive and cannot be economically justified except for larger systems, particularly those which are capable of supporting more than one user simultaneously. There are now a number of cheaper and simpler alternatives to hard disks, known collectively as floppy disks. These are available in a wide range of sizes and formats. As the name suggests, floppy disks are made of flexible material. The amount of information which can be stored on one floppy disk depends upon its diameter, the efficiency of the operating system, the recording density and whether or not both sides of the disk are used, however it is always substantially less than that which can be stored on a hard disk. The big advantage of disks over the other type of magnetic medium, tape, is that every part of the disk is equally accessible. This is possible because all computers using disks have a Disk Operating System (DOS). The disk operating system is a program which is used to format organise) disks and establish a directory on each disk. The directory contains information about the position of each separate block of information on the disk so that the read/write head can be immediately positioned over the relevant block. More detailed aspects of disk organisation and usage will be covered in Chapter 8.

If tape is used as backing store, access to information must be sequential, since the tape must be wound through from the start of the tape to the required block of information. Inevitably this means that access to information stored on tape is much slower than for information stored on disk. Many of the microcomputers currently available use domestic audio tape recorders and tape cassettes as inexpensive backing store devices. However, a consequence of this is that information transfer rates from such cassettes is exceedingly slow when compared with disks. Whereas a long program could be transferred from disk to RAM in a few seconds, the same process may take up to ten minutes if a domestic tape recorder is used.

1.4.5 Output devices

A computer is only useful if the results of its information processing can be passed on to the user. This is not an automatic process and will only occur if the computer has been instructed to deliver the information and, in addition, the necessary equipment, or output device, is connected to the system. The type of output device used will depend upon the circumstances but by far the commonest method is to display the information on a Cathode Ray Tube (CRT) in either a monitor or a domestic television set. These two devices, although very similar in appearance, are not identical and special circuits must be provided in the computer to convert the normal output of video signals into a form which domestic television sets can use. If a CRT is used then the output device is usually referred to as a VDU (Visual Display Unit).

The other main output device of computer systems is an electronic printer. Printers are used to produce hard copy, as opposed to the soft copy found on a VDU. There is an enormous range of printers available. These differ in both mechanism and price, but in general those which produce a printed output of a quality similar to that of electronic typewriters are the most expensive. The designers of computer systems must provide special circuits which interface the processor to the outside world as represented by the output devices. The methodology used is essentially a reverse of that employed with the input devices, in that a PIA converts the electronic signals used by the CPU into a format suitable for the output device selected.

1.5 Summary

You are probably now suffering from a surfeit of terminology and a feeling that you will never understand computers. Obviously computer systems are very complex pieces of equipment but a detailed knowledge of the principles of operation is not necessary even for advanced programming. The only way to gain familiarity with computers and the associated terminology is to use them. There is no substitute for 'hands on' experience. You cannot physically damage a computer with any programming commands, no matter how incorrect they are. You may on the other hand destroy any information that is currently in memory which may mean starting again!

Chapter 2 Programming Languages

2.1 Introduction

When a computer is first switched on it will probably respond with a message such as 'READY OK'. In the absence of any further actions from the user it will remain in an inert condition more or less indefinitely. This is because the computer will only perform tasks as a consequence of the execution of a set of instructions which are collectively known as a computer program. Indeed the sign-on message provided by a computer is produced by a program which is already resident in the computer's memory. This program is responsible for the normal operations of the computer and is therefore referred to as the operating system or monitor. Operating systems are discussed in more detail in Chapter 8.

A computer program is a series of instructions which are held either permanently or temporarily in the computer's memory. When the computer user issues the command to begin execution of a program the operating system will ensure that the first instruction in the program is carried out, followed by the second, third, etc. Programming is the technique of arranging the appropriate instructions in such a manner that the computer will carry out the task required by the programmer.

Programming actually consists of three distinct steps:

1. identify the problem (this can be difficult);

2. devise a solution to the problem and express this solution as a series of steps; this solution is known as the ALGORITHM;

3. convert the algorithm into a sequence of instructions using the selected programming language; this stage is known as coding and it produces the program.

The following example is an illustration of this procedure.

Example 2.1

1. Problem: what is the mean (average) of five numbers?

2. Solution: the algorithm
    a) obtain the values of the five numbers
    b) add them up
    c) divide the total by five
    d) print out the result.

3. Coding: coversion of the algorithm into the BASIC programming language. The following is not an elegant conversion but uses simple instructions which should be meaningful to the novice.

```
10 INPUT A,B,C,D,E
20 LET SUM = A + B + C + D + E
30 LET MEAN = SUM / 5
40 PRINT " THE MEAN IS ";MEAN
50 END
```

The first two steps in programming do not require any computing knowledge, therefore all that is preventing you from writing successful programs is a lack of knowledge about programming languages.

The instructions which form the basis for any computer program must be presented to the computer in an unambiguous format which is not easily achieved with normal human language. A series of programming languages have gradually evolved over the last thirty years. They started as lists of binary numbers that the computer recognised as particular instructions but have gradually evolved into languages which bear a resemblance to English. The following sections will introduce the concept of programming languages by considering initially the only language that any computer recognises, its own.

2.2 Machine language programs

Each of the many microprocessors that form such a fundamental part of all computers is capable of performing several hundred different operations. These operations fall into three main categories:

a) simple arithmetic on two numbers
b) manipulation of information stored in the memory
c) initiation of alternative pathways through a program depending upon the result of a comparison between two numbers.

The complete list of operations available to a CPU is known as the instruction set. This is fixed for each type of microprocessor. Differences between instruction sets are partially responsible for some of the differences between computers. Each of the instructions is identified by a reference number, the OPCODE (OPeration CODE), and a program consists of a list of opcodes plus associated items of data arranged in a sequence that will perform the required task. A program which is written using these opcodes or a derivative of them is known as a MACHINE LANGUAGE program since it utilises the computer's innate language. These are difficult languages to master because they consist entirely of numbers. Fortunately most biologists will not need to master this type of language! However, for those of you who are curious, a very simple example of machine language programming is included below. This example is written in the language of the 6809 microprocessor and illustrates how two numbers could be added together and the resulting sum stored for future use.

Example 2.2

1. Problem: what is the sum of two numbers, 215 and 33?

2. Solution:
   a) put the number 215 into register A of the 6809 CPU;
   b) add 33 to the 215 already in register A;
   c) store the resulting sum in a suitable memory location.

3. Coding: conversion into 6809 machine language.
   Before this conversion can be completed we need to know the opcodes for three instructions. Placing a number into register A is accomplished by the instruction whose opcode is 134; adding a number to the value currently in register A is instruction 139. The final operation, storing the resulting sum, is achieved by calling up instruction 151.

The process is outlined in Example 2.3 and it is assumed that in this computer the program and associated data items are stored in memory locations 1 to 6 and that the sum will be placed in memory location 200.

Example 2.3

Memory location	*Number stored*	*Notes*
1	134	Opcode for load register A
2	215	The number to be loaded into register A
3	139	Opcode for add the following number to the existing contents of register A
4	33	The number to be added
5	151	Opcode for the instruction which stores the contents of register A into the memory location to be specified
6	200	The memory location where the sum will be stored

The program consists, therefore, of the sequence of numbers: 134 215 139 33 151 200.

How does the computer distinguish between those numbers which are opcodes and those which are items of data? A computer has no inherent means for doing this; instead it is implied by the structure of the opcodes. The computer assumes that the first number in the sequence is an opcode. If the instruction identified by that opcode requires a number to operate on then the computer will assume that the number, or its location in memory, will be contained in the next memory location. It is, therefore, very important that program execution begins at the appropriate memory location. In normal circumstances this will be determined by the operating system but it is possible, particularly when using machine language programming, that execution will begin at the incorrect place in memory. If this happens you will soon realise that computers are not intelligent! The computer will attempt to follow a sequence of instructions which are now probably meaningless. For example, if the computer had started program execution at memory location 2 it would have attempted to carry out the instruction whose opcode is 215 and it would no longer add 215 to 33. If the above program had been successful and we then examined the contents of the computer's memory, we would have found that locations 1 to 6 were unchanged while location 200 now contained the number 248 (215 + 33). Machine language programming is, therefore, very difficult, but there are few alternatives if speed of execution is absolutely essential.

Machine language programming can be made slightly easier if assembly language is used. Assembly languages replace the opcodes with mnemonic equivalents. These mnemonics are translated into machine language opcodes by a translator program called an assembler. Example 2.4 is the assembly language equivalent of Example 2.3.

Example 2.4

Instruction	Explanation
LDA#215	Load register A with the number 215
ADDA#33	Add 33 to the existing contents of register A
STA#200	Store the contents of register A in memory location 200

2.3 High level languages

Most people find it difficult to program in machine language. Therefore, it is fortunate that there are much simpler alternatives which are more than adequate for most biological problems. Special programs have been written, in machine language, which allow the user to communicate with computers in languages which approximate to English. These special programs, which most users are not even aware of, work by translating 'English' commands into machine language instructions. These translation programs are more commonly referred to as high level languages. The first of these languages appeared in the late 1950's and they have been proliferating ever since. Most biologists will neither require nor encounter the majority of these languages. The most common of the high level languages, due to its almost universal implementation on microcomputers, is BASIC (an acronym from Beginners All-purpose Symbolic Instruction Code). This language is adequate but not necessarily ideal for many biological applications. If your only contact with computers is via one of the many microcomputers now available then you may not come into contact with other high level languages. However, if you have access to a larger mini- or mainframe computer it is probable that a range of languages will be available to you, many of these will be more efficient than BASIC. The next section will introduce a selected range of high level programming languages.

BASIC
This language was developed in the United States at Dartmouth College in the late 1960's. It was intended that it would make computers available to everyone, and to a large extent it has succeeded in this aim, particularly since the boom in microcomputers which started in the late 1970's. BASIC, unlike many of the earlier high level languages, was not written especially for mathematicians or engineers. The structure of earlier high level languages reflects the mathematical tasks for which they were developed.

Consequently, people from other disciplines often find them difficult to master. BASIC has a further big advantage over these earlier languages in that each part of the program can be tested as it is written. Most of the early languages and their derivatives are known as compiled languages. In a compiled language the complete program is written in pseudo-English and only then converted, using a program known as a compiler, into machine language. Compiled languages produce programs which are generally very fast in execution because they are operating directly in the language of the machine. However, the process of compilation is time consuming and frustrating if your original program does not work (this is quite common!). However, once a program has been successfully compiled it remains in that form and is not translated every time that the program is executed.

BASIC is an interpreted language which means that the instructions are converted into machine language every time that the program is run. It is possible, with an interpreted language, to check each separate instruction as it is entered. The penalty for this ease of use is an increase in the time needed to a run a program because of the delay produced by the conversion process. It is interesting to note that BASIC compilers are beginning to appear for many microcomputers. These compilers produce compiled, i.e. machine language, versions of BASIC programs and thereby remove one of the main problems of this popular language. Computer scientists and professional programmers often criticise BASIC both because it is slow and because it is relatively easy to write functional, but sloppy and confusing, programs. There is a body of opinion that some of this criticism arises because BASIC has removed the monopoly once enjoyed by these professionals!

Since BASIC is the language to which most people will have access, the majority of this book will be concerned with an explanation of this language using biological problems as examples. However, there is a problem with BASIC which results directly from its success. Computer manufacturers have developed varieties (dialects) of BASIC which are unique to their machines. A result of this is that it is seldom possible to directly transfer BASIC programs between different computers. The examples which will be presented later in this book use a collection of BASIC commands which should work on most computers. Inevitably this will mean that facilities, which some users take for granted, will not be covered because they are not universal. However, the examples should provide programs that the reader could improve and customise by using any special facilities offered by the model they are using.

Pascal
Pascal was developed in 1970 by Professor Wirth of Zurich. He named the language after the famous mathematician Blaise Pascal. It is a language which is popular with computer scientists because of its 'structured' nature, and has been identified by many people as the language most likely to replace BASIC. However, although it has many advantages over BASIC, it is a harder language for computer novices to master. The biggest obstacle against its general adoption is that it is generally not available on microcomputers as a standard language and, therefore, most people will be unable to experiment with it. This is unfortunate because there is little doubt that Pascal is a much better programming language than its older competitor, BASIC. It now seems unlikely that Pascal will oust BASIC from its

dominance on microcomputers. Because of this, and the fact that very little of the current commercial software used by biologists is written in Pascal, it is unlikely that many biologists will need a working knowledge of it.

COBOL

COBOL is another acronym. It is derived from COmmon Business Oriented Language. As its name suggests, this language is not intended for scientific usage. It was devised for the American Department of Defense who required a language to help with problems such as stock control. The main characteristics of COBOL are concerned with the handling of datafiles and the construction and printing of forms. It is unlikely that biologists will have any need for COBOL unless they become administrators!

FORTRAN

This is an important language which is frequently used for the solution of complex mathematical problems. The name was derived from a description of its use, FORmula TRANslation. FORTRAN and BASIC are reasonably similar because BASIC was developed from FORTRAN. Biologists will probably find FORTRAN easier to learn as a second language following BASIC. It has programming structures which are advantageous for advanced mathematical procedures and many of the important advanced statistical packages used by biologists are written in FORTRAN. Fortunately, in most instances the only FORTRAN knowledge required by a biologist is an understanding of how data is structured in this language. However, if your work is likely to involve long and complex mathematical manipulations, then it is probably advisable that you gain some knowledge of FORTRAN, particularly if speed of computation is important.

FORTH

This is a language which may become popular on microcomputers because of its availability and price. There is, therefore, a good probability that biologists will come into contact with it. FORTH has a surprising origin for a microcomputer language. It was developed in America in the early 1970's by Charles Moore in order to overcome some of the deficiencies of FORTRAN. He used the language to control a large astronomical telescope. The name FORTH is an abbreviation for FOURTH! Moore considered his new language belonged to the fourth generation of computer languages but his computer allowed him to use names of five letters only!

FORTH is a much harder language to master than BASIC but it is potentially very useful, if only because of its speed of execution. In particular FORTH would appear to have a future as an equipment control language because it is easier to use than machine languages, while still retaining much of their speed. One of the most interesting features of FORTH is that the user can, by writing programs, extend the available facilities. The language consists of a dictionary of words, which are, in effect, programs. Ideally the word should reflect its programming function. New words are added to the dictionary by defining them in terms of existing words. In FORTH it would be possible to define a new word such as VARIANCE, which when called up would calculate the variance of a set of numbers. Once defined VARIANCE could then be used to define new words.

FORTH is a difficult language to master because it uses two unusual (at least to biologists) concepts. These concepts are Reverse Polish Notation

(RPN) and stack manipulation. RPN is a method of representing calculations which presents the components of a calculation in a different format from that normally encountered. Calculations usually have two components: the OPERANDS or numbers and the OPERATORS, for example + and −. While the usual form of representation is to place the operator between the operands, RPN places the operator after the operands. In RPN 7 + 8 would become 7 8 + and 17 − 2 would be represented as 17 2 −. The stack is a temporary memory store which operates on the LIFO principle. This is yet another acronym, Last In First Out, and it describes the way in which the stack functions. The usual analogy used to explain LIFO is a stack of plates: the last plate to be placed on the stack would be on top and therefore would be the first to be removed; the same is true for numbers in the memory stack. Programming in FORTH is certainly more difficult than BASIC but it does present a challenge which is satisfying to overcome.

This review of high level languages was not intended to be comprehensive. There are many other programming languages available, particularly on larger computers. Most of these have specific applications, for example LISP is a LISt Processing language which is used in Artificial Intelligence (AI) work. If you wish to find out more about the range and applications of the many other high level languages a good source of preliminary information is the 'better' popular computing magazines such as *Practical Computing*, *Personal Computer World* and *Byte*. The bibliography (Appendix E) contains a list of books which should be useful as an introduction to various computing languages.

Chapter 3 Introduction to BASIC

This chapter will introduce some of the commands of the BASIC programming language and will explain how a program may be developed by following a sequence of logical steps. These procedures will be illustrated by developing a program for a particular biological problem.

3.1 Program development

The development of any computer program consists of two main stages: planning and coding. Planning involves thinking through the problem and then deciding how it should be tackled, i.e. the algorithm is developed. The second stage, coding, is the process of converting the algorithm into commands of the programming language. Initially, you will probably find that this second stage is the hardest and most time consuming. This is because you will be unfamiliar with the commands and will, inevitably, be programming relatively simple problems. As you become more familiar with the commands there is a danger that the first stage will become neglected. However, this must not be allowed to occur, because planning is the most important part of any program development. If the algorithm is carefully thought through then coding should become a relatively rapid process. Unfortunately, the structure of BASIC allows the user to skip the planning stage more or less completely and yet still produce programs which give the correct answers. However, programs which have been written without the planning stage usually take longer to refine and can only be described as bad programs.

A good computer program should, as a minimum, fulfill the following criteria.

a) It should give the correct responses over the entire range of possible values for all variables used.
b) A user of the program should be presented with unambiguous options, and incorrect responses by the user should be detected and rejected wherever possible.
c) Anyone familiar with the programming language should be able to understand what the program is doing. This also applies to the author.
d) The program should be as efficient as possible.

Most of these objectives can be achieved if the practices of structured programming are adopted. In essence this means breaking down the problem into a number of more or less independent steps or blocks which can be processed separately. In languages such as Pascal, which are block structured, this process is more or less automatic, but in BASIC, which is not block structured, the author of the program must make a conscious effort to achieve this aim.

3.2 BASIC programs

BASIC is a line orientated language. This means that the program instructions are organised as a sequential list with each instruction occupying a single line prefixed by a number. The smallest line number is at the start of the program. The line numbers are then incremented at a rate determined by the programmer. When the program is executed (run) the instruction occupying the lowest numbered line is the first to be carried out. This is followed by the next highest numbered instruction and so on. BASIC is, therefore, a linear language. However, it is possible to break this linear sequence and jump to any line number in a program by using a variety of BASIC commands. This facility has a number of important and useful consequences which will be discussed later. It also allows the programmer to write very confusing programs which can be very difficult to correct or modify.

Even if the problem has been carefully thought through the program may still not work correctly. When this occurs the program is said to contain errors or bugs. These can be of two types, logic errors and syntax errors. If the bug is a logic error then the program may function in an apparently normal manner but may produce an incorrect answer. These errors are usually the result of faulty algorithms and can only be corrected by rethinking the algorithm. Syntax errors are probably the most common bugs and arise when incorrectly constructed commands are present in the program, usually as a result of typing errors. Syntax errors are removed by editing the program and correcting the faulty instruction.

Before we can begin to write programs it is necessary to understand something about the ways in which a computer may be controlled. The commands which are made available to you by the BASIC interpreter fall into two categories, system commands and program commands. System commands look after the general operation of the computer and are not normally used in programs. The actual commands available to you will depend upon your computer, the two most useful and universal being RUN and LIST. RUN instructs the computer to begin execution of the BASIC program which is currently present in the memory. LIST instructs the computer to provide a copy of that program on the current output device such as the VDU or printer. Program commands are those which are used within programs although some of them will operate outside of a program in what is called the direct execution mode.

The RETURN key, which may also be called the ENTER or NEWLINE key, is exceedingly important. When material is typed at a keyboard it is held in a buffer until the RETURN key is pressed. The computer will only be able to respond to that material after it has left the buffer. Consequently, if you type RUN or LIST but do not press the RETURN key nothing will happen because the command has not yet been passed on to the CPU. Similarly, when you are typing in lines of BASIC they will be held in the buffer until you press the RETURN key. This key should therefore be pressed at the end of each line of BASIC. The buffer has a limited size which is usually 255 characters. If you attempt to type more than this number of characters without pressing the RETURN key then the display will appear to freeze up because there is no room left in the buffer.

Another useful feature of BASIC is that programs may be used interactively. In some programming languages this may not be possible because the user has no access to the program while it is running. BASIC, however, allows the user to interact with the program while it is running. This means that the computer responds immediately to any data items that are entered into the program. This kind of user control over the operation of a computer program is also known as real time operation.

The biological problem to be considered in the remainder of this chapter involves a relatively trivial equation which is encountered by most biologists at least once in their careers. If you have access to a computer which supports the BASIC language it will help you if you type in the programs exactly as seen in the following sections. As you become more confident do not be afraid to experiment with the programs. Many of your experiments will fail but this can be very useful, since finding out the reason for the failure will teach you a great deal about BASIC.

3.3 The problem

We want to write a program that will tell us the relative centrifugal field (RCF) produced by a spinning centrifuge. The RCF can be calculated using a simple equation which is ideal for conversion into a simple program.

$$\text{RCF} = \frac{4 \cdot \pi^2 \cdot (\text{revs min}^{-1})^2 \cdot \text{radius of rotor arm}}{3600 \times 981}$$

This problem can be broken down into a number of simple steps which can then be coded as separate blocks. Initially the blocks can be identified by a series of questions.

1. What information should be given to the user at the start of the program?

2. How can the values for the variables be entered into the program?

3. How can the equation be represented in BASIC?

4. What information should be given to the user at the end of the program?

The program will therefore consist of four blocks:

    information to the user, an introduction
    entry of values for the variables
    calculation of RCF
    information to the user, a summary.

These blocks can now be coded into BASIC commands.

3.4 The program structure

3.4.1 Block 1: Introduction

It is good practice to include in a program details of exactly what it will be doing. This will be particularly beneficial if other people are going to use the program but it will also remind you if you use it several months after it was completed. This first block will, therefore, present the user with information about the program.

```
10 REM           CALCULATION OF RELATIVE CENTRIFUGAL FIELD
20 REM
30 REM ------BLOCK ONE------INFORMATION -------------------------
40 PRINT "THIS PROGRAM CALCULATES THE RELATIVE CENTRIFUGAL FIELD"
50 PRINT "PRODUCED BY CENTRIFUGATION."
60 PRINT
70 PRINT "THE EQUATION USED IS :-"
80 PRINT "     RCF = 4 . PI^2 . (REVS MIN)^2 . RADIUS"
90 PRINT "         ------------------------------"
100 PRINT "                   3600 . 981"
110 PRINT
120 REM           END OF BLOCK ONE
```

The program lines have been numbered in increments of ten but there is nothing special about this particular interval. Line numbers 1, 2, 3, 4, ..., 12 or 3, 6, 9, 12, ..., 36 or 2, 5, 6, 10, 15, ..., 87 would all have been equally valid. It is simply that it has become an accepted convention that increments of ten should be used. It is also useful in that it allows the insertion of up to nine new program lines between any of the existing lines. However, if you have thought out the problem before you started the coding then you should not need to add additional lines!

Two BASIC commands have been used in this first block and they are REM and PRINT. Any program line which begins with the command REM is effectively ignored by the computer. REM is BASIC shorthand for REMark and is included so that the program may contain internal documentation. Line 10 explains what the complete program will be doing, line 20 is merely decorative in that it improves the look of the program listing and lines 30 and 120 mark the beginning and end of the block. If lines 10, 20, 30 and 120 were removed from the program it would have no effect on the operations to be carried out.

The PRINT command forces the computer to direct an output to the current output device. It will be assumed from this point on that the output device is a VDU. Everything that is enclosed between the quotation marks on lines 40, 50, 70, 80, 90 and 100 will be printed exactly as seen. Lines 60 and 110 result in the printing of a blank line. This is because, unless otherwise directed, the PRINT command is always terminated by a carriage return and line feed, that is the next character to be printed will be at the start of the next lowest line. Note that it is not possible to use superscripts or special symbols such as π in a program so that π^2 must be represented by PI^2. The ^ symbol (it may be ↑ or even [on some machines) is a BASIC command which represents the instruction 'raise to the power of' and PI^2 should be interpreted as PI raised to the power of 2 or π^2.

The program block consisting of lines 10 to 120 is, in fact, a complete program which could be successfully executed. If the command to initiate program execution, which is usually RUN, was issued, then the output from the above program should resemble the display shown below, although there may be minor differences between computers. You will have problems with some of these lines if your screen display is less than 64 characters wide. You should amend the PRINT statements, possibly by splitting a long PRINT statement into two lines.

```
>RUN

THIS PROGRAM CALCULATES THE RELATIVE CENTRIFUGAL FIELD
PRODUCED BY CENTRIFUGATION.

THE EQUATION USED IS :-
        RCF = 4 . PI^2 . (REVS MIN)^2 . RADIUS
              --------------------------------
                      3600 . 981

OK
>
```

3.4.2 Block 2: Data entry

The entry of values into the program is a more complex operation than those encountered in the first block. Two problems must be considered. How should the variables be represented in the program? How can values be assigned to these variables?

Representation of variables
Initially the discussion of variables in BASIC will be restricted to those which can have only numeric values. Variables are represented by names or symbols similar to those encountered in algebra. Most people will be familiar with equations such as the straight line equation shown below.

$$y = a + bx$$

This equation contains four variables represented by the single letters a, b, y and x. If values are substituted for a, b and x then the value of y can be calculated. Changing the values of a, b and x will not change the symbols but will result in a new value for y. BASIC allows variables to be represented in exactly the same way. However the letters A–Z would restrict us to only 26 possible variables in any one program. While this may seem a reasonably large number it is quite possible for even relatively simple programs to use more than 26 variables. Consequently all dialects of BASIC allow the use of more extended variable names, but it is here that we come into contact with the first problem produced by the range of dialects. All dialects will allow variable names composed of a single letter followed by a number between 0 and 9 to give variable names such as A1, B7, F3 and Z9. Many dialects will allow the use of variable names made up of two letters (for example WW, DF, PW) and even complete words such as SAMPLE, LOCATION1, COUNT, SUM. The most useful variable names are those of the latter type since they vastly improve the legibility of a program. However, since such names are not accepted by all dialects they will not be used in most of the subsequent programs even though their use is highly recommended if the facility is available. Variable names will be restricted to those which are valid in all dialects, single letters or a single letter followed by a single digit.

If a BASIC dialect allows the more complex variable names to be used there are two common problems which may arise. Often a dialect will allow long names to be entered in the program but it will only use a set number of characters for its own internal reference system. If variable names are restricted to two letters, any variable names which begin with the same two letters will be assumed to be the same; for example, the interpreter would not distinguish between two variables labelled COUNT and CORE since both begin with the letters CO. Many BASIC dialects will not allow variable names to be used which contain reserved BASIC command words. Thus a variable labelled IMPRINT would not be allowed since it contains the BASIC command PRINT. TOTAL cannot be used because it contains the reserved word TO (see p. 36).

In the equation for the calculation of RCF there are three variables, RCF, radius and speed of rotation, and four constants, 4, pi, 981 and 3600. The numbers 4, 981 and 3600 do not need to be represented by variable names although there are circumstances where it could be advantageous (see Chapter 14). Four variable names must be used to represent RCF, radius, speed and pi. Any four names which would be accepted by the BASIC interpreter could be used but it is good practice to use names which are as meaningful as possible. Radius will therefore be represented by R, speed by S, pi by P and RCF by F. (If longer names are allowed then obviously the names RADIUS, SPEED and PI would be more useful.)

Assigning values to variables

Three methods are available for assigning values to variables in programs. Each method uses a different BASIC command: LET, INPUT, or READ ... DATA.

The LET command can be used to assign values, so that in the present example a statement such as

```
LET P = 3.1416
```

could be used. This program statement should be translated as 'let the variable P take the value of 3.1416'. In this context the sign = does not mean equals, as it used to assign a value to a variable. Some other programming languages try to avoid this confusion by using the symbol := to assign values.

Values could also be assigned to R and S in exactly the same way, so that block 2 would look like:

```
130 REM ------BLOCK TWO-----ASSIGNMENT OF VALUES TO VARIABLES-----
140 REM       R = RADIUS, S = SPEED AND P = PI
150 LET R = 15
160 LET S = 3000
170 LET P = 3.1416
180 REM       END OF BLOCK TWO
```

Although this would work correctly the program can only calculate RCF for a centrifuge having a radius of 15 cm and a speed of rotation of 3000 revs min^{-1}. New values for R and S can only be entered if lines 150 and 160 are retyped. If the program is to be of any value it should be capable of greater flexibility in that it should calculate RCF for a range of values of R and S. This flexibility can be produced by using INPUT or READ ... DATA.

INPUT can only be used if the program is being used interactively, i.e. its inclusion will mean that the user will need to provide certain information while the program is running. The format of this command is INPUT variable name, for example INPUT R or INPUT S. When the computer executes this command the program will halt and a prompt symbol will appear on the output device, which indicates to the user that the computer is waiting for a value to be entered via the input device. The type of prompt symbol used will depend upon the computer but is usually a question mark or something similar. The value of the variable is entered at the input device, usually a keyboard, and after the RETURN key is pressed this value is passed to the program and assigned to the variable contained in the INPUT statement. If no extra information is presented to the user of a program then the execution of an INPUT statement will result in the appearance of just the prompt symbol. Will the user understand what is required? The usual answer to that question is no! Consequently, the program must be made more 'user friendly', or in other words it should tell the user what is expected. This can be achieved by preceding the INPUT statement with a PRINT statement that tells the user what is required. If INPUT statements are used, block 2 could be rewritten as follows to provide a large degree of flexibility.

```
130 REM -----BLOCK TWO-----ASSIGNMENT OF VALUES TO VARIABLES--------
140 PRINT "ENTER THE RADIUS OF THE ROTOR ARM"
150 INPUT R
160 PRINT "ENTER THE SPEED OF ROTATION"
170 INPUT S
180 LET P = 3.1416
190 REM   END OF BLOCK TWO
```

In the above sub-program the only lines which are absolutely necessary for its operation are 150, 170 and 180, since it is at these points that the values are assigned to the variables. Line numbers 130 and 190 provide internal documentation by labelling the start, function and end of the block. Line numbers 140 and 160 present the user with information and instructions when the program is running. This block could be checked by instructing the computer to RUN 130, thus ignoring lines 10–120 and beginning execution at line 130. (Unfortunately this is a command which is not implented in all dialects of BASIC. The BBC, MZ-80K and ATARI microcomputers do not allow this form of the RUN command. It can be simulated by entering a new line as follows: 1 GOTO Linenumber (where Linenumber is replaced by the number required, e.g. 1 GOTO 130). The GOTO command is explained later. The output produced by such an operation would resemble:

```
>RUN 130

ENTER THE RADIUS OF THE ROTOR ARM
? 15                    (the 15 was entered by the user)
ENTER THE SPEED OF ROTATION
? 3000                  (again entered by the user)

OK
>
```

The INPUT command is the most useful way of getting small amounts of data into programs. It loses its attraction when a large number of data items are to be entered, since such activities invariably result in wrong values being entered. This may not be a problem in sophisticated programs which have error traps and data editing facilities built in, but more usually it will mean starting again from the beginning. The most useful way of getting large quantities of data into programs is to use a pair of BASIC commands, READ and DATA. (There are more sophisticated methods available which use data files but these will not be covered until Chapter 8.)

The format of the READ command is READ variable name, for example, READ R, READ T1, READ X. The value which will be read into the variable is contained in a DATA statement, for example, DATA 3, DATA 15, DATA 3.142. These two linked commands do not need to be adjacent to each other because of the way in which they are used by the BASIC interpreter. When a READ command is encountered the interpreter will search through the program for the first item of data which follows a DATA statement and this will be entered into the variable following that READ command. The next READ statement will result in the next item of data being used. In fact the computer keeps track of DATA statements by using a pointer. A pointer is a number which identifies where in memory the next DATA item can be found. As each piece of data is read the pointer is incremented to take account of this fact, thus ensuring that each piece of data is only used once. Each DATA statement can be used for several items if they are separated by commas. However, the sequence of numbers must be that required by the READ statements. Block 2 could be written as follows if READ/DATA commands are used to enter the values required by the program.

```
130 REM -----BLOCK TWO-----ASSIGNMENT OF VALUES TO VARIABLES----------
140 READ R
150 READ S
160 READ P
170 DATA 15,3000,3.1416
180 REM     END OF BLOCK TWO
```

If this section was run by again typing RUN 130 then the screen display would be as follows.

```
>RUN 130

OK
>
```

Apparently nothing has happened. This is because this block contains no PRINT commands and therefore nothing has been directed to the VDU. However, the block will still have fulfilled its function in that the variable R would now equal 15, S would equal 3000 and P would be 3.1416. This can be verified by using the PRINT command in direct command mode. Enter PRINT R,S,P and press the RETURN key to see the current values of these three variables. In fact the block could have been shorter since several variables can be attached to one READ command. Line 140 could have been written as READ R,S,P and it would have performed exactly the same function as the previous lines 140–160. If you make this change to line 140 and type RUN 130 you will find that the computer will issue an error message to tell you that it is out of data. This is because lines 150 and 160 still contain READ statements but there is no data remaining in the program that can be used. This error can be corrected by removing lines 150 and 160. There are two methods which can be used to bring this about. If you type 150 and then press RETURN, you will have overwritten the previous line 150 with a blank line which the interpreter will then ignore. The second method is to use a system command, DELETE (or DEL on some models), to remove the unwanted lines. You must specify which lines you wish to delete by typing DELETE 150, or more efficiently, DELETE 150–160. (This is another command which has various formats. The BBC micro achieves the same result with DELETE 150, 160. If in doubt check the manual!)

When DATA statements are used the data is contained within the program. A disadvantage of this is that new values can only be entered by editing or retyping the lines containing the data. It is, therefore, an inefficient system to use when relatively few items of data are required but very useful when large sets of data are being used. This is because if a mistake has been made in one of the data items the complete set need not be re-entered; instead the DATA line carrying the mistake can be edited and the correct value entered. DATA statements also provide a useful way of entering test data into a program while it is being developed. If test data is held in DATA statements it need not be re-entered each time that the program is run.

Since the centrifugation problem only requires two different values for each run of the program the most appropriate method for entering these values is by the use of INPUT commands.

3.4.3 Block 3: The calculation

Coding of the RCF equation requires a knowledge of how BASIC handles arithmetic operations. BASIC is provided with five arithmetic operators as shown below.

Symbol	Operation
$\wedge$	exponentiation (raising a number to a power)
*	multiplication
/	division
+	addition
−	subtraction

28

BASIC computations are very similar to algebraic expressions. The only major difficulty is with multiplications. Algebraic multiplications such as xy for x multiplied by y must not be written in BASIC as XY. BASIC would recognise XY as a variable in those dialects which permit two or more letter variable names. Others would not recognise it and the program would crash. The xy multiplication must be coded as X $*$ Y. Several simple examples are shown below.

Algebraic expression	BASIC statement
$x = a + b$	LET X = A + B
$x = a + b - c$	LET X = A + B − C
$x = z/y$	LET X = Z/Y
$x = a^2 - b$	LET X = A^2 − B
$x = ab + c$	LET X = A $*$ B + C

An added complication is that BASIC does not always evaluate expressions from left to right. This is because the arithmetic operators have different levels of precedence. Exponentiation has the highest precedence and will be evaluated first. This is followed by multiplication and division which are of equal precedence. Addition and subtraction occupy the lowest level of precedence and are, therefore, evaluated last. If two operators of equal precedence occur in the same BASIC statement they are evaluated on a left to right basis. The existence of precedence means that some algebraic expressions must be carefully coded if they are to produce the desired result. The following examples should illustrate this point.

Let $a = 10$, $b = 8$ and $c = 2$

Algebraic		BASIC	
$\dfrac{a}{b + c}$	= 1.0	LET X = A/B + C	X = 3.25
$a + \left[\dfrac{b}{c}\right]^2$	= 26	LET X = A + B/C^2	X = 12
$a.b/c$	= 40	LET X = A $*$ B/C	X = 40
$\dfrac{a - b}{c}$	= 1.0	LET X = A − B/C	X = 6

The above equations can be evaluated correctly by using parentheses () to force a new order of precedence, expressions within parentheses being evaluated before all others. It is possible to nest parentheses, e.g. B^2 $*$ (A $*$ (N −1)); in such cases it is always the expression contained within the innermost pair that is evaluated first. In the example given the sequence of operations would be N−1, multiplied by A. B would then be squared before the final multiplication was carried out. The previous examples are shown again below, but this time parentheses have been used to force the desired sequence of operations.

Algebraic		BASIC	
$\dfrac{a}{b + c}$	= 1.0	LET X = A/(B + C)	X = 1.0
$a + \left[\dfrac{b}{c}\right]^2$	= 26	LET X = A + (B/C)^2	X = 26
$a.b/c$	= 40	no change required since left to right precedence gives the correct result	
$\dfrac{a - b}{c}$	= 1.0	LET X = (A − B)/C	X = 1.0

Consider the RCF equation:

$$RCF = \frac{4 \times \pi^2 \times (\text{revs min}^{-1})^2 \times \text{radius}}{3600 \times 981}$$

The sequence of operations should be:

1. square π

2. square the revs min^{-1}

3. multiply the products of (1) and (2)

4. multiply the product from (3) by 4.0

5. multiply the product in (4) by the radius.

This will give us the numerator.

6. multiply 3600 by 981.

This will provide the denominator.

7. divide the numerator in (5) by the denominator in (6).

The numerator can be calculated as follows:

$$4 * P^\wedge 2 * S^\wedge 2 * R$$

sequence of operations   3   1   2   4

This will provide the correct value for the numerator. The only problem is that the denominator must be evaluated before the division of the numerator. This can only be achieved if the operation 3600 × 981 is enclosed within parentheses to give the BASIC expression:

LET F = 4 $*$ P$^\wedge$2 $*$ S$^\wedge$2 $*$ R/(3600 $*$ 981)

Note that without the parentheses the numerator would be divided by 3600 and the result multiplied by 981.

Block 3 becomes:

```
200 REM ------BLOCK THREE------CALCULATION OF RCF----------------
210 LET F = 4 * P^2 * S^2 * R / (3600 * 981)
220 REM       END OF BLOCK THREE
```

3.4.4 Output of results

All that remains to complete this first program is to output the result to the user. When line 210 has been executed the value of the RCF, for the values used, is held by a variable F. This value will only be printed if the program contains the appropriate instructions. The computer does not know that this is the value you are seeking! Some suitable program lines are shown below.

```
230 REM -------BLOCK FOUR----RESULTS SUMMARY--------------------
240 PRINT "RELATIVE CENTRIFUGAL FIELD IS"
250 PRINT F
260 REM         END OF BLOCK FOUR
```

3.4.5 The end!

There is one more statement that should be included in all programs. This is an END statement. One reason for including it is that it may make you think a little more about the structure of your program when you are deciding where it should be placed.

```
270 END
```

3.5 The complete program

```
10 REM ---------CALCULATION OF RELATIVE CENTRIFUGAL FIELD------------
20 REM
30 REM ---------BLOCK ONE------INFORMATION-------------------------
40 PRINT "THIS PROGRAM CALCULATES THE RELATIVE CENTRIFUGAL FIELD"
50 PRINT "PRODUCED BY CENTRIFUGATION."
60 PRINT
70 PRINT "THE EQUATION USED IS :- "
80 PRINT " RCF = 4 . PI^2 . (REVS MIN)^2 . RADIUS"
90 PRINT "       --------------------------------"
100 PRINT "                  3600 . 981"
110 PRINT
120 REM END OF BLOCK ONE
130 REM ---------BLOCK TWO------ASSIGNMENT OF VALUES TO VARIABLES------
140 PRINT "ENTER THE RADIUS OF THE ROTOR ARM "
150 INPUT R
160 PRINT "ENTER THE SPEED OF ROTATION "
170 INPUT S
180 LET P = 3.1416
190 REM END OF BLOCK TWO
200 REM ---------BLOCK THREE----CALCULATION OF RCF--------------------
210 LET F = 4 * P^2 * S^2 * R / ( 3600 * 981 )
220 REM END OF BLOCK THREE
230 REM ---------BLOCK FOUR-----SUMMARY------------------------------
240 PRINT "RELATIVE CENTRIFUGAL FIELD IS "
250 PRINT F
260 REM END OF BLOCK FOUR
270 END
```

3.5.1 A sample run of the program

```
>RUN

THIS PROGRAM CALCULATES THE RELATIVE CENTRIFUGAL FIELD
PRODUCED BY CENTRIFUGATION.
```

```
THE EQUATION USED IS :-
    RCF = 4 . PI^2 . (REVS MIN)^2 . RADIUS
          --------------------------------
                     3600 . 981

ENTER THE RADIUS OF THE ROTOR ARM
? 15
ENTER THE SPEED OF ROTATION
? 3000
THE RELATIVE CENTRIFUGAL FIELD IS
1509.12088

>OK
```

It is important to realise at this early stage that computers work to set levels of numeric precision. This has important consequences whenever calculations are performed on numbers which have a large number of significant figures. Even though an answer may be produced which has eight decimal figures it does not mean that the answer is correct to that level of precision. This important problem is discussed further in Appendix A.

3.6 Summary

The following BASIC statements have been introduced in this chapter.

REM — An abbreviation for REMark which is used to allow internal documentation in programs. Nothing which follows a REM statement is used by the BASIC interpreter.

PRINT — PRINT produces an output to the current output device which consists of all characters enclosed within double quotes ("). If PRINT is followed by a variable name then the current value of this variable is printed.

Variable names — Numeric quantities may be stored within programs as variables. These may consist of a single letter or a letter plus a number. Note that many dialects allow the use of more complex variable names ranging from two to long strings of letters.

LET — This command is used to assign values to variables. The value may be numeric or another variable name, e.g. LET X = 2.3 ; LET X = Y. Note that the symbol = does not mean equals in this context. In many dialects the LET command is optional and may be omitted.

INPUT — When programs are being run interactively this command can be used to assign values to variables. When an INPUT command is executed the BASIC interpreter will send a prompt symbol to the output device. The program will halt at this point until a number is entered and the RETURN key is pressed.

READ ... DATA   These commands cannot be used independently of each other. A READ command is used to assign a value to a variable. The value is contained within the program following a DATA statement. The two statements do not need to be adjacent to each other. Once a value has been read from a DATA statement it is not used again.

DELETE (DEL)   This is a system command. It is used to remove unwanted lines from programs. Since it is a system command it cannot be used within a program (normally).

Arithmetic   There are five arithmetic operators available in BASIC.
operators   They are: $\wedge$ exponentiation; $*$ multiplication; / division; + addition; $-$ subtraction. When more than one operator occurs in the same statement they are evaluated in their order of precedence. The highest precedence is exponentiation, next is multiplication and division (equal precedence) and the lowest order is held by both addition and subtraction. If two operators have equal precedence they are evaluated from left to right. Parentheses () are used to force a different order of precedence since all expressions contained within parentheses are evaluated first.

END   This is used to mark the end of the program. In most dialects it is optional.

3.7 Problems

You should now attempt to write programs to use with the following problems. Before starting to code the equations think carefully about the structure of the program. Appendix D contains sample programs and printouts.

1. Write a program which will give the dry weight of plant material when it is known from previous work that dry weight is equal to 0.653 of the wet weight.

2. Write a program which can be used to determine the output of the left ventricle of the heart ($1 \ min^{-1}$). The output can be calculated from the following equation.

$$output = \frac{oxygen \ consumption \ (ml \ min^{-1})}{[oxygen_{art}] - [oxygen_{ven}]}$$

Some suitable values are: oxygen consumption $= 250 \ ml \ min^{-1}$; arterial oxygen concentration $= 190 \ ml \ l^{-1}$; venous oxygen concentration $= 140 \ ml \ l^{-1}$.

3. The surface area of a human can be estimated by the following equation.

$$S \text{ (surface area m}^2) = 0.007184 \times W^{0.452} \times H^{0.725}$$

where W is the weight in kg and H is the height in cm.

Write a program which can be used to estimate the surface area of a human.

4. The following equation can be used to determine the amount of infra red radiation (Wm^{-2}) emitted by the sky.

$$IR_{sky} = (0.4 + 0.06 \sqrt{e_w}) \, \sigma \, T_a^4$$

where e_w = humidity in millibars
T_a = air temperature in degrees Kelvin (°C + 273)
σ = Stefan-Boltzmann constant 5.57 x 10^{-8} Wm^{-2} K^{-4}

Square roots can be obtained by raising a number to a power of 0.5.
Determine the amounts of infra red radiation for the following values.

temperature °C	e_w (saturation vapour pressure mbar)
0	6.11
10	12.27
20	23.37
30	42.43

Chapter 4 Loops and Arrays

One of the facilities which makes computers so useful is their ability to carry out repetitive operations without making mistakes or becoming bored. This is possible because all programming languages contain at least one set of instructions which can control repetitive functions. This is achieved by establishing loops in a program. A loop is a segment of program that is executed several times in succession. This chapter will consider only those commands which establish unconditional loops, that is loops which are executed for a fixed, predetermined number of times. Students often have difficulty with the concept and operation of loops. Frequently this is because they are looking for difficulties which are not there!

4.1 The simple FOR ... NEXT loop

Consider the problems involved in the addition of a long list of numbers. If this task is attempted by humans it is quite common to obtain as many different totals as there were attempts at the addition! Such a problem would seem to be an ideal task for a computer. A BASIC program which could carry out such a task would follow quite closely the method used by humans. However, humans carry out simple operations such as these without consciously thinking about what they are doing. Therefore, before we construct a program, we should consider exactly the steps which are followed in the addition of a column of numbers.

$$
\begin{array}{r}
5 \\
8 \\
3 \\
7 \\
\underline{9} \\
\end{array}
$$
Total ?

The technique employed by most people would look something like:

$$
\begin{array}{rcl}
5 + 8 &=& 13 \\
13 + 3 &=& 16 \\
16 + 7 &=& 23 \\
\underline{23 + 9} &=& \underline{32} \\
\text{Total} &=& 32 \\
\end{array}
$$

The addition has been completed by carrying out a series of additions where the next number in the list has been added to a running total. The little example above misses out two steps which have been carried out subconsciously. What was the initial running total, and how did you know when you got to the end of the list? Computers cannot operate subconsciously and therefore they will need to be provided with the information contained in the answers to these two questions. The initial running total was zero which became 5 after the first addtion, $0 + 5 = 5$, had been carried out. The end of the list was recognised because there were no more numbers below the 9. This is not an operation which computers can perform easily, it must be built into the program. One of the simplest methods of doing this is to include in the program information about the length of the list. A summary of the operations carried out in the addition of a list of numbers is shown below.

a) Initialise (give a variable an initial value) the running total to zero.
b) Add the next value in the list to the running total.
c) Repeat operation (b) as many times as there are numbers in the list.
d) The final value of the running total is the sum of the numbers.

The steps (a) to (d) are now sufficiently well defined that they can be used as an algorithm for the addition of a list of numbers. The coding of this algorithm will use a pair of BASIC commands, FOR and NEXT. FOR ... NEXT statements are used to establish unconditional loops in a program which will be executed for a fixed number of times as determined by the format of the FOR statement.

A computer is only able to execute a loop for a fixed number of occasions because it keeps track of how many times the loop has been completed. This is possible because each FOR ... NEXT loop has an associated INDEX VARIABLE which is used for this purpose. There is, also, additional information which cannot be assumed but must be provided by the program. This information relates to the initial and final values of the index variable but also includes information about the rate at which it should be incremented. In the simple addition problem the initial value will be 1, i.e. start with the first number in the list, and the final value for the index variable will be equal to the length of the list. The index value will be incremented by 1 on each passage of the loop, so that each number in the list is added to the total. Do not worry if, at the present, you are unable to understand why the index variable should ever be incremented at a rate other than 1. The value of this facility will be demonstrated later.

The format of a BASIC FOR ... NEXT loop is shown below.

```
10 FOR I = 1 TO 5 STEP 1
 .
 .                          program instructions to be repeated
 .
50 NEXT I
```

In this example line 10 marks the start of the loop and line 50 marks the end of the loop. All instructions on lines numbered between 10 and 50 would be carried out as many times as the loop is repeated. The instructions on line 10 should be translated as 'For I, the index variable, takes an initial value of 1 to a final value of 5, with an incremental step of 1'. The NEXT I instruction on line 50 will result in the current value of the index variable being incremented at a rate defined by the STEP statement. This probably sounds very complicated and is best illustrated by following through all of the changes which would occur during execution of the FOR ... NEXT loop shown above.

When the instructions on line 10 are first executed several variables are initialised (this is achieved automatically using the values contained within the FOR command). In the example above the index variable I would be set to its initial value of 1. The index variable can be any legal variable name but by convention the letters I, J and K are used most frequently. If a value other than 1 had been present in the FOR statement (e.g. FOR I = 3 TO 10) then the index variable would have been initialised to that value. The CPU stores the step and end values in memory locations reserved for this purpose. In most dialects of BASIC a step value of 1 need not be specified since in the absence of a step value a default value of 1 is assumed. Line 10 could, therefore, have been shortened to:

```
10 FOR I = 1 TO 5
```

After the initialisation on line 10 had been completed, the statements on lines between 10 and 50 would be executed. Control would then pass to line 50 where the instruction NEXT I would be executed. At this point the index variable I would be incremented by the amount specified in the STEP statement. In the example above the STEP value is 1 so I would now become 1 (present value) + 1 (step value) = 2. The new value of the index variable is then compared with the specified end value and if the new value is less than or equal to the end value then control is returned to the line immediately following that containing the FOR statement. These intermediate lines are again executed until control returns to line 50. The index value is again incremented and compared to the end value. After the five loops of the example have been completed I would equal 5 and execution of line 50 for the fifth time would raise I to a value of 6. Since this is now greater than the specified end value of 5 the loop has been completed the required number of times. Therefore the loop is terminated and the next line of instructions to be executed would be that which immediately follows the line containing the NEXT instruction.

A very common mistake made by novice and experienced programmers is to misread or mistype I as a 1 and vice versa.

The execution of simple FOR ... NEXT loops is best demonstrated by a series of short programs. If you have access to a computer try several variations on the examples shown below. Try to predict what the output will be before you run the programs.

Example 4.1

Program	Output	Notes
10 FOR I = 1 TO 5	> RUN	start the program
20 PRINT I	1	line 20 1st time
30 NEXT I	2	line 20 2nd time
40 PRINT I	3	line 20 3rd time

```
50 END                          4        line 20 4th time
                                5        line 20 5th time
                                6        line 40
                              > OK       line 50 end
```

Note that in the absence of a STEP value on line 10 the default value of +1 was adopted.

Example 4.2

Program	Output	Notes
`10 FOR J = 2 TO 4 STEP .5`	`> RUN`	start the program
`20 PRINT J`	`2`	line 20 1st time
`30 NEXT J`	`2.5`	note the increment of 0.5
`40 PRINT J`	`3.0`	
`50 END`	`3.5`	
	`4.0`	line 20 4th time
	`4.5`	line 40, J is now greater than
	`> OK`	4, the specified end value

It is also possible to have loops which count backwards. This is possible if the STEP value is negative. Each time that the NEXT statement is executed the index variable is decremented by the specified step amount. If the index value is then less than the specified end value the loop is terminated. Example 4.3 demonstrates this facility.

Example 4.3

Program	Output	Notes
`10 FOR W = 3 TO 1 STEP -1`	`> RUN`	start the program
`20 PRINT W`	`3`	line 20 1st time
`30 NEXT W`	`2`	line 20 2nd time
`40 PRINT W`	`1`	line 20 3rd time
	`0`	line 40, note that W is
	`> OK`	now less than the end value

Note that if a negative STEP was specified but the start value for the loop was less than the end value (e.g. FOR K = 1 TO 3 STEP −1) the loop would normally be completed once. This is because the index variable is only compared with the end value when the NEXT statement is executed. Therefore the intermediate statements will have been executed once.

The three examples above all used numbers to specify the loop parameters. However, BASIC provides considerable flexibility with loops since the parameters can be set by using variables. The statement

```
FOR J = S TO E STEP I
```

would be valid if values had previously been assigned to the variables S, E and I.

In Chapter 1 an example of a BASIC program was given which could add five numbers and calculate their mean. This is not a very useful program since it could only be used with groups of five numbers. A much more flexible program can be written if a FOR ... NEXT loop is included to code the algorithm given earlier in this chapter.

Example 4.4

```
 10 REM ADDITION AND MEAN CALCULATION
 20 LET N = 0
 21 LET T = 0
 22 LET X = 0
 30 PRINT "HOW MANY NUMBERS IN THE LIST"
 40 INPUT N
 50 PRINT "ENTER THE NUMBERS NOW"
 60 FOR I = 1 TO N
 70 INPUT X
 80 LET T = T + X
 90 NEXT I
100 PRINT " TOTAL IS "; T
110 PRINT " MEAN IS "; T/N
120 END
```

Line 10 is a description of the program. Since it follows a REM statement it is not executed. Lines 20 to 22 initialise the three variables, N, T and X, to a value of zero. Line 30 produces an output to the screen requesting the user to enter the number of values to be added. This number is entered in response to the prompt symbol produced when line 40 is executed. Whatever value is entered will be stored by the computer as a variable labelled N. Line 50 provides more information to the user by requesting that the numbers to be added should now be entered. Line 60 marks the start of a loop. It will be indexed by a variable called I which will start with an initial value of 1. I will then be incremented by 1 each time line 90 is executed. The end value for the loop is set as N and can therefore be assigned different values each time the program is run.

Line 70 will produce a request for an input. The number entered will be stored as a variable called X. Since line 70 is within the loop it will be executed several times. On each occasion the existing value of X will be overwritten by the new value. Consequently, when the loop is terminated, X will carry the last value entered. All other values will have been lost. Line 80 is interesting because of its algebraic nonsense! It is, however, a very valuable BASIC statement which is used in a large number of programs. It must not be read as 'Let T equal T plus X'. The correct expression is 'let the variable T take the value of the current value of T plus the value of X'. In this way the sum of the column of numbers is gradually accumulated as the variable T. When the loop has been executed N times T will equal the sum of all N numbers.

Since lines 100–120 are outside the loop they will only be executed once. Line 100 is obviously a PRINT statement but it is constructed differently from those used before. The symbols contained between the double quotes (") will be printed exactly as seen in the program. The semicolon (;) is a print control statement. This suppresses the carriage return which normally follows a PRINT statement, and the next character to be printed will be placed immediately after the words TOTAL IS. Since the letter T is not enclosed within quotes it is not treated in the same way as the T's in TOTAL. The computer recognises it as a variable and therefore prints the value of the variable T rather than the letter. If the sum of the numbers was 275, line 100 would produce the output: TOTAL IS 275. Line 110 is executed in exactly the same way as line 100 except that the expression T/N is first evaluated and the result printed following the words MEAN IS.

The following is an example of what would be seen if the above program was executed.

```
>RUN

HOW MANY NUMBERS IN THE LIST
? 5
ENTER THE NUMBERS
? 5
? 8
? 3
? 7
? 9
TOTAL IS 32
MEAN IS 6.4

> OK
```

This simple program can form the basis for many quite complex programs where the accumulation of a large number of values is required. However, I would hope that none of the readers of this book will ever need to use a computer to add up five numbers! Biologists should not be that innumerate.

4.2 Numeric arrays

Although the last program can be used to add up long lists of numbers, it has the disadvantage that when the addition has been completed only the last number entered is retained in memory (as the variable X). BASIC provides a means by which all the values could be retained, without resorting to the use of a different variable name for each value. This method involves the use of a numeric array which is simply a list of numbers in which each number can be identified by its position in the list. In the example above an array of five numbers was summed. These numbers can be described by their positions in the array. 5 is the first, 8 is the second, 3 is the third, etc. In algebraic notation they would be identified by subscripts, $x_1, x_2, x_3 .... x_n$. In mathematics the symbol Σ means 'the sum of' and the total of a column of numbers is represented as:

$$\sum_{i=1}^{n} x_i$$

This should be read as 'sum all of the values of x starting at the first value ($i = 1$) and ending at the nth value'. This sounds very like the algorithm that was used in the previous section, except that subscripts are now used to identify individual values in the list. BASIC also allows the use of subscripted variables which are represented as VARIABLE NAME (SUBSCRIPT), for example X(2), W(19), S1(12). The addition problem could therefore be written as LET T = X(1) + X(2) + X(3) + X(4) + X(5). This is, however, no better than the A + B + C + D + E method. Fortunately the subscript does not need to be a number. It can be a variable name, for example X(I), W(J), S1(V). Consider what happens to the value of I in a loop with a step value of 1 and an initial value of 1. It is incremented by one on each passage of the loop and will therefore be 1 on the first loop so that X(I) would be X(1), on the second loop I would be equal to 2 and therefore X(I) would now be X(2), etc. The FOR ... NEXT loop in the addition program could therefore be written as follows.

```
60 FOR I = 1 TO N
70 INPUT X(I)
80 LET T = T + X(I)
90 NEXT I
```

Since I has a different value each time the loop is completed lines 70 and 80 will refer to different subscripted variables, as shown below.

Loop number	Value of I	Variable used in lines 70 and 80
1	1	X(1)
2	2	X(2)
3	3	X(3)
4	4	X(4)
.	.	.
N	N	X(N)

It is important to note that, even if they occurred in the same program, a variable called X would have no connection with a subscripted variable X(I). This is because simple and subscripted variables are not treated in the same way by computers.

Space must be reserved in memory for variable arrays. This is achieved with the DIM command. DIM is an abbreviation for DIMension and the process by which memory space is reserved is called dimensioning. DIM statements are usually placed near the start of the program and take the form of: DIM VARIABLE NAME(NUMBER OF ARRAY ELEMENTS). The following would all be valid statements: DIM X(100); DIM D1(6); DIM N(35). Computers may differ in the ways in which arrays are treated but the following points are valid for most models.

a) If an array has not been dimensioned but the program contains subscripted variables, the BASIC interpreter will normally reserve space for eleven elements. Note that the BBC microcomputer BASIC requires all arrays to be dimensioned.

b) The first element in the array may be numbered with the subscript zero. Therefore, the eleven element array in (a) is numbered from 0 to 10.

c) It is usually possible to dimension an array with a variable. This is known as Dynamic Dimensioning, for example DIM X(N).

d) It is not usually possible to redimension an array while a program is running. This would involve a large scale reorganisation of material in the memory.

The following program can be used to calculate several simple statistics. They are calculated by using the equations shown below.

sample size $= n$ (number of values)
mean $= $ total$/n$
sum of squares $= \Sigma x^2 - (\Sigma x)^2/n$
sample variance $= $ sum of squares$/n-1$
standard deviation $= \sqrt{\text{variance}}$

The program is split into five blocks.

1. Initialisation

2. Title and user information

3. Data entry

4. Calculation of statistics

5. Output of results

The main problems to be overcome in this program are associated with the calculation of the sum of squares, variance and standard deviation. However, as shown above, these three quantities are related and their calculation requires the accumulation of two totals. The first is the sum of the squared values of x, i.e. $x_1^2 + x_2^2 + x_3^2 ... + x_n^2$. The second total is the sum of x squared, i.e. $(x_1 + x_2 + x_3 ... + x_n)^2$. Note that these two quantities are not equal. The sum of squared values of x can be obtained by squaring each value of x and adding it to an accumulating total. The sum of the x values will be obtained in a similar manner.

```
10 REM SIMPLE STATISTICS PROGRAM
20 REM VARIABLES D = STANDARD DEVIATION, M = MEAN, N = SAMPLE SIZE
21 REM S = SUM OF SQUARES, X(I) = ARRAY OF VALUES
22 REM V = VARIANCE, T = TOTAL, T2 = SUM OF SQUARED VALUES
30 REM ------------BLOCK ONE-----INITIALISATION--------------------
31 LET D = 0
32 LET M = 0
33 LET N = 0
34 LET S = 0
35 LET V = 0
36 LET T = 0
37 DIM X(50)
38 REM NOTE CHANGE THE SIZE OF THE X ARRAY FOR SAMPLES LARGER THAN 50
40 REM ------------BLOCK TWO-----USER INFORMATION-------------------
50 PRINT "CALCULATION OF SIMPLE STATISTICS"
51 PRINT
60 PRINT "THIS PROGRAM WILL CALCULATE THE MEAN, STANDARD DEVIATION AND"
70 PRINT "VARIANCE FOR A SAMPLE OF UPTO 50 NUMBERS"
```

```
 71 PRINT
 80 REM ------------BLOCK THREE---DATA ENTRY------------------------
 90 PRINT "HOW LARGE IS THE SAMPLE ";
100 INPUT N
120 PRINT "ENTER " ; N ; " NUMBERS "
130 FOR I = 1 TO N
140 PRINT I;"   ";
150 INPUT X(I)
160 NEXT I
170 REM ------------BLOCK FOUR----CALCULATIONS---------------------
180 FOR I = 1 TO N
190 LET T = T + X(I)
200 LET T2 = T2 + X(I)^2
210 NEXT I
220 LET M = T / N
230 LET S = T2 - T^2 / N
240 LET V = S / (N - 1)
250 LET D = V ^ 0.5
260 REM ------------BLOCK FIVE----RESULTS OUTPUT--------------------
270 PRINT
280 PRINT "SAMPLE SIZE        = " ; N
290 PRINT "MEAN               = " ; M
300 PRINT "VARIANCE           = " ; V
310 PRINT "STANDARD DEVIATION = " ; D
320 END
```

Lines 10–22 provide internal documentation for the program by identifying its function and the variables used. They are, of course, not essential for the operation of the program.

Block 1 is used to initialise the variables to zero. Although this process is unnecessary in most dialects of BASIC it is good programming practice. It will help you with other languages, such as FORTRAN and Pascal, in which variables must be declared at the start of a program. Also, in long programs it can help to speed up program execution (see Chapter 14 for details). The DIMension statement on line 37 is essential if the program is to be used for samples larger than 11 numbers. If samples larger than 50 are to be used then this line must be altered to accommodate the extra memory requirements.

Block 2 presents the user with information about the program. The appearance of the program could be improved by the addition of a statement which would clear the display before presenting this information. There are a number of dialect-specific commands which will clear the display, the two most common being CLS and HOME. If such a command is available it should be included between lines 40 and 50.

The third block is concerned with data entry. Line 90 asks the user for the size of the sample. Note that it is terminated with a semicolon outside the double quotes. This will suppress the carriage return which would usually follow the execution of a PRINT statement. Consequently, the prompt which is produced by the INPUT statement on line 100 will follow the question. A comparison of the screen displays with and without the terminal semicolons is shown below.

Program	Output
90 PRINT "HOW LARGE IS THE SAMPLE "	HOW LARGE IS THE SAMPLE
100 INPUT N	?
90 PRINT "HOW LARGE IS THE SAMPLE " ;	HOW LARGE IS THE SAMPLE ?
100 INPUT N	

The presentation of line 120 is similarly improved by embedding the variable N within the PRINT statement. If N was equal to 25 line 120 would look like:

```
ENTER 25 NUMBERS
```

Lines 130 to 160 form a loop whose parameters are set by the format of line 130. The statements on lines 140 and 150 will be repeated N times. I will begin with a value of 1 and will be incremented by 1 until after N loops it will be equal to N + 1. At this point I will be greater than N, the end value for the loop. Consequently, control will pass to the first line after the loop, which is line 170. Lines 140 and 150 have been structured to improve the screen presentation. Line 140 will print the current value of I. The semicolon will suppress the carriage return and two spaces (" ") will be printed. Since this is also followed by a semicolon the carriage return will again be suppressed. The result of this is that the prompt produced by the input statement on line 150 will occur on the same line as the value of I. In this way the user can keep track of his position in a list of numbers.

Block 4 is concerned with the various calculations which are required. Another loop is established between lines 180 to 210. This loop is used to accumulate the totals of x and x^2 (T and T2). Note that these two values could have been obtained by including the statements within the first loop. However, it is advisable to keep the different components of a program separate from each other. Although debugging (correcting) would be relatively simple in a short program such as this, it can be difficult in long, complex programs. Separation of the different blocks aids this procedure and it is never too early to develop desirable techniques.

The four statistics are calculated on lines 220 to 230. Note the difference between T2, a variable, and the expression T^2 on line 230. Line 240 contains a good demonstration of the need to use parentheses to force a desired sequence of mathematical operations. Without the parentheses the calculation on line 240 would have been S divided by N with 1 being subtracted from the result. The square root of V is obtained by raising it to the power of 0.5.

The final block is used to output the results to the user. Again, note the use of semicolons to improve the presentation. Line 320, END, is strictly outside Block 5. Although not usually essential it is good practice to include an END statement.

Some programmers suggest that the computational algorithm used in Example 4.5 will only be valid for numbers which are neither very large nor very small. This is because the accumulation of the squared values of X may produce a number with more significant digits than the computer can handle. This could result in rounding errors as described in Appendix A. An alternative computational algorithm is described in Example 4.6 and problem 5 at the end of this chapter.

A sample run of this program is shown below.

```
>RUN
CALCULATION OF SIMPLE STATISTICS

THIS PROGRAM WILL CALCULATE THE MEAN, STANDARD DEVIATION AND
VARIANCE FOR A SAMPLE OF UPTO 50 NUMBERS

HOW LARGE IS THE SAMPLE ?10
ENTER 10 NUMBERS
1   ?12.4
2   ?13.2
3   ?10.9
4   ?9.6
5   ?15.3
6   ?8.8
7   ?12.2
8   ?14.1
9   ?9.8
10  ?17.2

SAMPLE SIZE         = 10
MEAN                = 12.35
VARIANCE            = 7.20055539
STANDARD DEVIATION  = 2.68338506

>OK
```

4.3 Nested FOR ... NEXT loops and two dimensional arrays

Consider what would happen if one FOR ... NEXT loop is enclosed within another FOR ... NEXT loop. This process is called nesting and an example is shown below.

```
10 FOR I = 1 TO 3
20 PRINT "I = ";I
30 FOR J = 1 TO 4
40 PRINT "   J = ";J
50 NEXT J
60 NEXT I
70 END
```

The outer loop has I as an index variable and will be executed three times (line 10). The inner loop uses J as the index variable and will be executed four times (line 30). However because the J loop is contained within the outer I loop the entire J loop will be executed three times. If the program is run the PRINT statements on lines 20 and 40 will demonstrate this feature.

Sample run

```
>RUN

I = 1
    J = 1 ⎫
    J = 2 ⎪
    J = 3 ⎬  first time through the J loop
    J = 4 ⎭
I = 2
    J = 1 ⎫
    J = 2 ⎪
    J = 3 ⎬  second time through the J loop
    J = 4 ⎭
I = 3
    J = 1 ⎫
    J = 2 ⎪
    J = 3 ⎬  third time through the J loop
    J = 4 ⎭

>OK
```

The ability to nest FOR ... NEXT loops is very useful. It allows for the multiple execution of repetitive processes. Perhaps more importantly it also facilitates the use of two or more dimensional variable arrays. A two dimensional array is one in which each variable is identified by two subscripts, e.g. X(1,3); R1(10,25); T(X,Y). The two subscripts may be thought of as coordinates in a table. A table of data is one in which a data item can be described in terms of its row and column position. This relationship is shown below.

		Column number							
		1	2	3	4	5	6	...	n
	1	1,1	2,1	3,1	4,1	5,1	6,1	...	n,1
	2	1,2	2,2	3,2	4,2	5,2	6,2	...	n,2
Row	3	1,3	2,3	3,3	4,3	5,3	6,3	...	n,3
number	4	1,4	2,4	3,4	4,4	5,4	6,4	...	n,4
	...	...	...	...	...	...	...	...	...
	n	1,n	2,n	3,n	4,n	5,n	6,n	...	n,n

An array of data in this format should be a familiar sight to most biologists. Note that by convention the first subscript describes the column while the second is used for the row. The ability to identify and manipulate variables within such tables is exceedingly important. If you compare the output of the simple nested loop program with those of the array subscripts above you may begin to see how double subscripted variables can be used within programs.

Example 4.6

This program can be used as the basis for quite complex statistical tests such as analysis of variance. In this example a table of data will be entered and then various simple operations will be applied to the data to produce the information required for the more advanced statistical procedures. A complete analysis of variance program is not included for several reasons, the most important being that statistical procedures such as this are easily abused. It is important that the user has a thorough understanding of the underlying theory before using such tests. An excellent account of biological statistics can be found in Zar's book (see Appendix E for details; also included in Appendix E are details of two books which combine statistical theory with the appropriate BASIC programs).

Consider the following experiment. Three isolates of a fungus are grown on the same nutrient medium. Each treatment is replicated five times. We wish to know if the yield (amount of fungus produced in mg) differs between varieties. The following results were obtained.

Replicate number	Isolate number 1	2	3
1	13	17	19
2	15	22	18
3	9	21	20
4	12	19	21
5	10	23	18

The BASIC program which can be used to read in the data and then calculate the treatment (isolate) means is quite simple and makes use of a number of nested FOR ... NEXT loops. The program also calculates the sum of squares for each treatment. These will be needed if the program is expanded to carry out more statistical tests.

```
10 REM EXAMPLE 4.6 APPLICATION OF NESTED LOOPS
11 REM
20 REM VARIABLES USED :
21 REM T          NUMBER OF TREATMENTS
22 REM T(I)       NUMBER OF REPLICATES PER TREATMENT
23 REM X(I,J)     INDIVIDUAL REPLICATE VALUES
24 REM M(I)       TREATMENT MEANS
25 REM T1(I)      SUM OF REPLICATES PER TREATMENT
26 REM T2(I)      SUM OF THE SQUARED REPLICATE VALUES FOR TREATMENT I
27 REM T1         SUM OF ALL THE REPLICATE VALUES { SUM OF ALL X(I,J)s }
28 REM
30 REM    ------------BLOCK ONE-------INITIALISATION--------------------
31 DIM X(5,25)
32 LET T = 0
33 LET T1 = 0
40 REM    ------------BLOCK TWO-------USER INFORMATION------------------
```

```
41 REM CLEAR THE SCREEN IF COMMAND IS AVAILABLE
42 PRINT "ENTRY OF DATA INTO A TABLE"
43 PRINT
44 PRINT "CALCULATION OF COLUMN TOTALS, MEANS AND SUMS OF SQUARES"
45 PRINT
60 REM      -----------BLOCK THREE-----DATA INPUT-----------------------
61 PRINT "HOW MANY TREATMENTS ARE THERE ";
62 INPUT T
63 PRINT "HOW MANY REPLICATES ARE THERE IN :-"
64 FOR I = 1 TO T
65 PRINT "TREATMENT ";I;"  ";
66 INPUT T(I)
67 NEXT I
68 PRINT
70 PRINT "ENTER THE DATA NOW"
71 FOR I = 1 TO T
72 PRINT " TREATMENT ";I
73 PRINT " REPLICATE "
74 FOR J = 1 TO T(I)
75 PRINT J ;"     ";
76 INPUT X(I,J)
77 NEXT J
78 NEXT I
79 PRINT
80 REM      -----------BLOCK FOUR------CALCULATIONS---------------------
81 FOR I = 1 TO T
82 FOR J = 1 TO T(I)
83 LET T1(I) = T1(I) + X(I,J)
84 NEXT J
85 LET T1 = T1 + T1(I)
86 LET M(I) = T1(I) / T(I)
87 NEXT I
90 REM CALCULATE THE SUMS OF SQUARES
91 FOR I = 1 TO T
92 FOR J = 1 TO T(I)
93 LET T2(I) = T2(I) + ( X(I,J) - M(I) )^2
94 NEXT J
95 NEXT I
110 REM      ----------BLOCK FIVE------USER SUMMARY---------------------
120 PRINT "TREATMENT","MEAN","TOTAL","SUM OF SQUARES"
121 FOR I = 1 TO T
122 PRINT I,M(I),T1(I),T2(I)
123 NEXT I
130 END
```

Sample run (using the fungal growth data)
The line numbers which are being executed will be shown enclosed within
< >. It is often possible to use a similar facility when debugging programs
by issuing a system command, TRACE (or TRON: TRace ON). This will
result in line numbers being printed as they are executed.

```
>RUN                                                     Line Number(s)
                                                         < 10 - 41 >
ENTRY OF DATA INTO A TABLE                               < 42 >
                                                         < 43 >
CALCULATION OF COLUMN TOTALS, MEANS AND SUMS OF SQUARES  < 44 >
                                                         < 45, 60 >
HOW MANY TREATMENTS ARE THERE ?3                         < 61 >
HOW MANY REPLICATES ARE THERE IN :-                      < 62 - 64 >
TREATMENT 1 ?5                                           < 65 - 67 >
TREATMENT 2 ?5                                           < 65 - 67 >
TREATMENT 3 ?5                                           < 65 - 67 >
                                                         < 68 >
ENTER THE DATA NOW                                       < 70, 71 >
TREATMENT 1                                              < 72 >
REPLICATE                                                < 73, 74 >
1     ?13                                                < 75 - 77 >
2     ?15                                                < 75 - 77 >
3     ?9                                                 < 75 - 77 >
4     ?12                                                < 75 - 77 >
5     ?10                                                < 75 - 77 >
TREATMENT 2                                              < 72 >
REPLICATE                                                < 73, 74 >
1     ?17                                                < 75 - 77 >
2     ?22                                                < 75 - 77 >
3     ?21                                                < 75 - 77 >
4     ?19                                                < 75 - 77 >
5     ?23                                                < 75 - 77 >
TREATMENT 3                                              < 72 >
REPLICATE                                                < 73, 74 >
1     ?19                                                < 75 - 77 >
2     ?18                                                < 75 - 77 >
3     ?20                                                < 75 - 77 >
4     ?21                                                < 75 - 77 >
5     ?18                                                < 75 - 78 >
                                                         < 79 - 110 >
TREATMENT   MEAN   TOTAL   SUM OF SQUARES                < 120, 121 >
1           11.8   59      22.8                          < 122, 123 >
2           20.4   102     23.2                          < 122, 123 >
3           19.2   96      6.8                           < 122, 123 >
                                                         < 130 >
>OK
```

The operations of the program up to line 60 do not contain anything which is new or difficult to understand. They will not, therefore, be discussed. Block 3, the data input block, makes use of one simple and one nested FOR ... NEXT loop. The first loop, lines 64 to 67, makes use of the variable T. This is the number of treatments. This first loop allows the user to specify how many replicates, T(I), there are in each of the T loops. Note the construction of lines 65 and 66. Line 65 causes the word TREATMENT to be printed followed by the current value of the loop counter I. The semicolon which follows I forces the next group of characters, two spaces, to be printed on the same line. The complete line is terminated by another semicolon which means that the carriage return is again suppressed. Therefore the prompt symbol produced by line 66 will be on the same line as the treatment number.

The nested loops are on lines 71 to 78. The outer I loop will be completed T times, which is three times in the example. Therefore the inner J loop will be executed three times. On each of these three occasions the number of cycles of the J loop will be controlled by the value of T(I). In this example $T(1) = T(2) = T(3) = 5$ and therefore the J loop will be completed 3 × 5 times. This can be verified by counting how many times line numbers 75 to 77 were executed. Line 75 is used to achieve the same effect as line 65.

Block 4 is the main part of the program and contains those instructions which carry out the calculations. The nested loops on lines 81 to 87 are used to sum the columns. Each column is summed in turn by the J loop on lines 82 to 84. Once these column totals have been obtained they are used to calculate the grand total (T1) and treatment means (M(I)). Note that these last two operations are outside the inner J loop but still within the outer I loop.

The latter part of block 4 is used to calculate the sum of squares for each column. Example 4.5 made use of a different computational algorithm to obtain a sum of squares which is based upon an algebraic rearrangement of the method used here. The sum of squares is defined as the sum of the squared deviations of each item from its mean. Consequently, this calculation requires a separate loop to be used after the means have been obtained. Although slower than the Example 4.5 algorithm, it is less likely to generate rounding errors.

The final block is concerned with presenting information to the user, the only unusual feature being the FOR ... NEXT loop on lines 121 to 123. This is used to display the results in a tabular format. The commas between the items in the print list separates the four quantities on the display.

4.4 Summary

Loops and arrays are very important components of most programs. Unfortunately they are also a facility which many people find difficult to understand. The examples used in this chapter have not made use of any of the complex manipulations which are possible with arrays. Some of these will be introduced in later chapters. The best method by which an understanding of loops may be gained is to experiment with short programs that are little more than FOR ... NEXT loops. Some suggestions are included in the problems below.

FOR ... NEXT This pair of commands is used to establish and control unconditional loops. The syntax of these commands is: FOR index variable = initial value TO end value. This statement sets the parameters for the loop. The end of the loop is marked by NEXT index variable.

; The semicolon can be used as a print control command. When it is placed at the end of a print statement it suppresses the carriage return and forces the next PRINT statement to continue from where the last one finished.

The comma is used as a print control command. When it is placed between items in a PRINT list it forces them to be printed in different print fields (which differ between computers). In this respect it is similar to the tab function on a typewriter.

4.5 Problems

1. Write a program to sum a list of numbers. Use READ DATA statements to enter the data into the program.

2. Include a facility in the above program which will allow the user to see the value of one of the elements in the array. (Hint — use a PRINT statement to ask the question and follow this with an INPUT statement which can be used to enter the array subscript.)

3. Modify the program from Problem 2 so that the user can change the value of a specified array element. After this has been done print the array to verify that the correct change has been made. In what situations could such a procedure be of great value?

4. Write a program to read in the following information and then print out the table exactly as seen below.

```
PROBLEM FOUR DATA
1  1   1   1    1
2  4   8  16   32
3  9  27  81  243
```

5. Alter the program listed in Example 4.5 to make use of the sum of squares algorithm used in Example 4.6. Example 4.5 currently calculates the sum of squares as follows:

S = sum of squared values of x − the square of the sum of x/n

The alternative algorithm is:

S = sum of [{ $X(I,J)$ − mean of treatment I } squared]

Compare the two algorithms by running each program with identical sets of data which should include one set composed of small numbers (< 0.1) and another containing large numbers (> 10000).

Chapter 5 Functions

5.1 Outline of BASIC functions

Many mathematical functions and procedures are used in biomathematics. Usually these are evaluated by complex mathematical procedures or by consulting tables. Fortunately BASIC is provided with a number of inbuilt functions which simplifies their use in biomathematical procedures. The general format of the mathematical functions in BASIC is: function name (argument expression). The argument expression can be a number, a variable name or an arithmetic expression which can be evaluated. In all three cases the value of the argument expression must be one which can be used with the function. If the argument expression is outside the valid range for the function then an error message will be produced and execution of the program will cease.

The most common functions provided by BASIC dialects are shown in Table 5.1.

Table 5.1 Common functions provided by BASIC dialects

Function	Argument requirements	Evaluated as
SIN(X)	X must be in RADIANS	the sine of X
COS(X)	X must be in RADIANS	the cosine of X
TAN(X)	X must be in RADIANS	the tangent of X
ATN(X)	X must be in RADIANS	the arctangent of X (in radians)
INT(X)	X must be a real number	greatest integer LESS than or equal to X (NEVER rounds up)
SGN(X)	X must be a real number	-1 if X is less than zero 0 if X is zero $+1$ if X is greater than zero
ABS(X)	X must be a real number	the absolute value of X
SQR(X)	X must be a positive number	the square root of X (same as $X^{\wedge}.5$)
EXP(X)	X must be a real number	e^X (2.718289 to the power X)
LOG(X)	X must be a positive number	the natural logarithm of X . Log_{10} X is obtained by LOG(X) $\times$ 0.301
RND(X)		differs between dialects. Returns a random number that lies between certain limits. See below.

BASIC does not contain tables from which these values are obtained. The interpreter evaluates inbuilt functions by calling up machine language programs which may use complex algorithms to obtain an estimate of the function. These estimates are frequently approximations and may differ slightly from those found in tables. This can be an important consideration if a very high level of accuracy is required from the computations.

The functions can only be used as part of an arithmetic expression or in a LET statement. A program line such as 10 SQR(35.8) would not be valid. It would need to be written as 10 LET X = SQR(35.8), where X could be replaced by any other valid variable name.

5.2 The use of BASIC functions in programs

Many of the functions can be used in a relatively straightforward manner. However, others do require a certain amount of care. The following section will outline the major problems associated with the use of some of the BASIC functions.

5.2.1 Trigonometric functions

The four trigonometric functions are relatively simple to use provided one remembers that X must be converted to RADIANS before it can be evaluated. If X has been measured in degrees it can be converted into radians by a simple formula which could be included in a program.

angle in degrees = (angle $\times$ π/180) radians

Example

$$55° = 55 \times 3.1416/180$$
$$= 0.96 \text{ radians}$$

However, it should be remembered that care must be taken when dealing with angles greater than 90° (π/2 radians). Trigonometric functions cannot be evaluated directly for angles greater than 90°. Also, the sign of the function must be taken into account. Table 5.2 lists the actions which should be taken.

Table 5.2 Use of trigonometric functions

Angle in degrees (radians)		Function evaluated	Sign of function
0 to 90	(0 to π/2)	function (angle)	all are positive
90 to 180	(π/2 to π)	function (180 − angle)	COS, TAN are negative, SIN is still positive
180 to 270	(π to 3π/2)	function (angle − 180)	COS, SIN are negative, TAN is positive
270 to 360	(3π/2 to 2π)	function (360 − angle)	SIN, TAN are negative, COS is positive

Example 5.1

How would the sines, cosines and tangents be obtained for an angle of 123°?

Function to be evaluated is 180 − 123 = 57°
This must be converted to radians:

57 $\times$ π/180 = 0.99484 radians

Therefore

sine (123)   =   SIN(0.99484) = 0.8387
cosine (123) = − COS(0.99484) = −0.5446
tangent (123) = − TAN(0.99484) = −1.54

Most dialects of BASIC include only the four trigonometric functions described earlier. Other trigonometric functions can be evaluated by using a little mathematical manipulation. Four of the most common derived trigonometric functions are shown below.

SECANT(X)   = 1 / COS(X)
COSECANT(X)   = 1 / SIN(X)
COTANGENT(X) = 1 / TAN(X)
ARCSIN(X)   = ATN(X / SQR(−X × X + 1))

5.2.2 Integer function

In most dialects of BASIC the argument which is used with the INT function must be within the range −32767 to +32767. The INT function must be used carefully since it does not always produce the expected result. In normal mathematical usage the integer of a number is obtained by rounding up to the next whole number all numbers with a fractional value of greater than 0.5. Those numbers whose fractional value is less than 0.5 are rounded down. The BASIC INT function does not operate in this way. It always rounds downwards. A comparison of the two operations is shown below.

Number	Integer	INT(X)
13.23	13	13
15.56	16	15
98.94	99	98
−15.76	−16	−16
−21.22	−21	−22

Note that particular care is needed with negative numbers. Some dialects have a function, FIX(X), which is very similar to INT. FIX(X) truncates a number to an integer by removing any numbers following the decimal point. It is therefore identical to the INT function for positive values of X. However, if X is negative it produces a different result since it does not round downwards. Only the fractional portion of the number is removed.

The INT function can be used to carry out a normal rounding off function if the argument is changed slightly. The following expression will round off an argument to its nearest integer number.

LET X = INT(Y + 0.5)

5.2.3 LOG function

Logarithms are mathematical functions which many biologists have some trouble in understanding. They are, however, very useful in biomathematics. The logarithm of a number is the power to which the base of the logarithm (commonly 10, 2 or e) must be raised to give that number.

Hence:

$\log_b x = y$ where b = base, x = number and y = logarithm

therefore

$b^y = x$

This relationship is demonstrated below for three different bases and $x = 2$.

$\log_{10} 2 = 0.3010$ $10^{0.301} = 2.00$
$\log_2 2 = 1.0000$ $2^{1.00} = 2.00$
$\log_e 2 = 0.6931$ $e^{0.6931} = 2.00$

The implementation of the LOG function in BASIC is potentially of great value. The three different logarithms which biologists are likely to encounter differ with respect to their base. Common logarithms use a base of 10, natural logarithms have e (approximately 2.71828) as the base. The third type are logarithms with a base of 2.

The LOG function in BASIC provides the natural logarithm, i.e. to the base e, of the argument. If a logarithm to another base is required then further manipulation is again required. The following equation can be used to derive the logarithm of a number to any base if natural logarithms are available.

$\log_b x$ = natural log of x/natural log of b

where x = argument
      b = desired base

Logarithms to base 10 (common logarithms) and base 2 can be evaluated by the following BASIC statements. X is the argument whose logarithm is required.

base 10   LET Y = LOG(X)/2.3026
base 2   LET Y = LOG(X)/0.6931

5.2.4 RND function

The ability of computers to generate pseudo-random numbers is very important for both games and simulations where chance is an important component. Computers generate random numbers by using complex algorithms that may differ between machines but which always involve the use of seed numbers. The seed number is that which the random number generating program uses to start the process. The way in which the random number generator is seeded can be very important. On some computers it is possible that a particular game or simulation will always use the same random numbers unless particular care is taken over the programming.

There are many differences between computers in the ways in which the RND function is implemented. However, two general classes can be recognised. In the first the argument can be any integer number within the permitted range (usually -32767 to $+32767$ for 8-bit computers). The random numbers generated by this function will be equally distributed between zero and the argument used, i.e. all values are equally probable. In the second class the argument may be either −1, 0 or +1. The random numbers generated by these dialects will be equally distributed between 0 and 1.

When a random number generator has been seeded with a particular number it will begin to produce a sequence of numbers that is always the same for that particular seed number. In dialects of the second type an argument of −1 will restart the sequence of random numbers from the beginning and it will produce exactly the same sequence of random numbers as before. An argument of 0 will result in the last random number generated being repeated. If +1 is used then the next number in the sequence (for the current seed) will be produced. A new seed can be used if commands such as RANDOM or RANDOMIZE are available. Since dialects of this type will only produce random numbers between 0 and 1, some programming manipulation is necessary if numbers within a wider range are required. The following FOR ... NEXT loop will produce integer random numbers within a specified range (0–30 in this example).

```
10 FOR I = 1 TO 10
20 LET X = INT ( RND(1) * 31 )
30 PRINT X
40 NEXT I
```

An explanation of line 20 is probably required. The RND(1) statement will result in the production of a pseudo-random number between 0 and 1. Multiplying this quantity by 31 will extend the range of possible values from 0 to 30. The INT function is used to convert the resulting number into an integer format. However, because of the way in which the INT function works, all numbers between 29.0000 and 29.9999 would be rounded down to 29. Thus the range of potential numbers is not 0 to 31 but 0 to 30. The above program can be written more simply in those dialects which generate random numbers outside the range 0 to 1 since the multiplication would obviously be unnecessary.

It will be help you to understand how the RND function works if you run the above program and then change some of the parameters and compare the outputs produced.

5.3 Example programs using BASIC functions

5.3.1 Write a program which can be used to calculate the pH of a dilute, aqueous salt solution

The equation used is:

$$pH = pK_w - 1/2\ pK_b + 1/2\ \log_{10} c$$

where $pK_w = 14$; $pK_b = pK_w - pK_a$; pK_a varies for different salts
c = salt concentration (mol dm^{-3})

Algorithm used:

1. Initialisation of the variables

2. User information

3. Input of values for variables

4. Calculation of pH

5. Summary for user.

Only those aspects of the algorithm which are new concepts will be covered in detail.

The only new programming steps required for this program involve the use of the LOG function. The equation makes use of the $\log_{10}$ of the salt concentration. The program could ask the user to input the information in this format, but it would be more useful if the concentration was converted within the progam.

```
10 REM    CALCULATION OF THE PH OF DILUTE SALT SOLUTION
20 REM --------BLOCK ONE-----INITIALISATION---------------------
25 REM A = PKA, B = PKB, C = CONC, P = PH
30 LET A = 0
31 LET B = 0
32 LET C = 0
33 LET P = 0
40 REM --------BLOCK TWO-----USER INFORMATION--------------------
50 REM          INCLUDE CLEAR SCREEN COMMAND HERE
60 PRINT "CALCULATION OF THE PH OF A DILUTE, AQUEOUS SALT SOLUTION"
65 PRINT
70 PRINT "YOU WILL NEED THE FOLLOWING INFORMATION"
72 PRINT "    1. SALT CONCENTRATION ( MOL PER CUBIC DECIMETER )"
74 PRINT "    2. PKA FOR THE SALT ( FROM TABLES )
76 PRINT
80 REM --------BLOCK THREE---DATA INPUT----------------------------
90 PRINT "WHAT IS THE SALT CONCENTRATION ";
95 INPUT C
100 PRINT "WHAT IS THE PKA FOR THE SALT ";
110 INPUT A
120 PRINT
130 REM -------BLOCK FOUR----CALCULATION OF PH--------------------
140 LET B = 14 - A
150 LET P = 14 - B / 2 + ( LOG ( C ) / 2.3026 ) / 2
160 REM -------BLOCK FIVE----USER SUMMARY--------------------------
170 PRINT "THE PH OF THE SALT SOLUTION IS "; P
180 END
```

Note that since only two values are needed they were entered via INPUT statements. The only difficult line in this program is line 150. If you refer to the initial equation you will see how it was derived. The pK of water is 14 so that can be present as a constant. The $\log_{10}$ of the concentration is required and therefore the natural log, obtained by LOG(C), was converted by dividing by 2.3026. Both pK_b and the logarithm of concentration must be divided by two. The parentheses around the LOG(C)/2.3026 statement are not absolutely necessary since the normal order of precedence, left to right, would produce the correct result. However, they do clarify the computation for anyone using the program. (Note that in the United States mol per cubic decimeter is normally written as mols per liter.)

Sample run

What is the pH of a 0.02 mol dm^{-3} aqueous solution of sodium proprionate? The pK_a of proprionic acid is 4.85

```
>RUN

CALCULATION OF THE PH OF A DILUTE, AQUEOUS SALT SOLUTION

YOU WILL NEED THE FOLLOWING INFORMATION
   1. SALT CONCENTRATION ( MOL PER CUBIC DECIMETER )
   2. PKA FOR THE SALT ( FROM TABLES )

WHAT IS THE SALT CONCENTRATION ?0.02
WHAT IS THE PKA OF THE SALT ?4.85

THE PH OF THE SALT SOLUTION IS 8.5755

>OK
```

5.3.2 Example 2

Consider how the pH program could be altered to provide a table which relates concentration to pH. There are various ways in which this could be achieved. The first solution that may occur (to you) is simply to run the program several times, each time entering a different salt concentration. There is a much more efficient method which makes use of the fact that STEP rates can be specified for FOR ... NEXT loops. If a FOR ... NEXT loop solution is used all that is needed is the range of concentrations and the incremental rate. This method is demonstrated in the following program.

```
10 REM  MODIFIED PH CALCULATION
20 REM --------BLOCK ONE----INITIALISATION----------------------------
21 REM A = PKA, L = LOWEST CONC, H = HIGHEST CONC, S = INCREMENTAL RATE
22 REM B = PKB, P = PH
23 LET A = 0
24 LET L = 0
25 LET H = 0
26 LET S = 0
27 LET B = 0
28 LET P = 0
30 REM --------BLOCK TWO----USER INFORMATION------------------------
40 PRINT "THIS PROGRAM WILL PRODUCE A TABLE WHICH RELATES"
45 PRINT "SALT CONCENTRATION TO PH. YOU WILL BE ASKED TO"
50 PRINT "PROVIDE THE PK OF THE SALT ( FROM TABLES )"
```

```
53 PRINT "ALL CONCENTRATIONS ARE IN MOL PER CUBIC DECIMETER"
55 PRINT
60 REM --------BLOCK THREE--DATA INPUT----------------------------------
70 PRINT "WHAT IS THE PKA OF THE SALT ";
75 INPUT A
80 PRINT "WHAT IS THE LOWEST SALT CONCENTRATION ";
85 INPUT L
90 PRINT "WHAT IS THE HIGHEST SALT CONCENTRATION ";
95 INPUT H
100 PRINT "WHAT IS THE INCREMENTAL RATE FOR THE TABLE ";
105 INPUT S
110 REM -------BLOCK FOUR---THE CALCULATION AND PRINTOUT---------------
115 PRINT
120 LET B = 14 - A
125 PRINT "CONC","PH"
130 FOR I = L TO H STEP S
140 LET P = 14 - B / 2 + ( LOG ( I ) / 2.3026 ) / 2
150 PRINT I,P
160 NEXT I
170 END
```

There are several points to note in this program. The PRINT statement on line 125 provides the heading for the table that will be produced by the FOR ... NEXT loop. If it had been included within the loop it would have been printed every time that the loop was completed. The comma between CONC and PH is a print control statement. It has the effect of moving the cursor across the screen to the start of the next print field (similar to a tab key on a typewriter). Line 130 sets up the parameters for the loop. L is starting concentration and H is the final one. The number of times that the loop is completed will depend upon the difference between H and L and the size of S, the STEP value.

The calculation of the pH is carried out by the expression on line 140. Note that the current value of I, the loop counter, is now used to provide the value of the concentration. The PRINT statement on line 150 will result in the current value of I being printed (this is the concentration used in the calculation). Also, on the same line, but separated because of the action of the comma, will be the pH for that concentration.

Sample run

```
>RUN

THIS PROGRAM WILL PRODUCE A TABLE WHICH RELATES
SALT CONCENTRATION TO PH. YOU WILL BE ASKED TO
PROVIDE THE PKA OF THE SALT ( FROM TABLES )
ALL CONCENTRATIONS ARE IN MOL PER CUBIC DECIMETER

WHAT IS THE PKA OF THE SALT ?4.85
WHAT IS THE LOWEST SALT CONCENTRATION ?.02
WHAT IS THE HIGHEST SALT CONCENTRATION ?.24
WHAT IS THE INCREMENTAL RATE FOR THE TABLE ?.04
```

```
CONC              PH
.02               8.5755205
.06               8.81407959
.1                8.92500324
.14               8.99806678
.18               9.05263867
.22               9.09621347

>OK
```

This short and simple program can provide a useful reference table from only a small amount of user supplied information. You may find it useful to experiment with this program by using different values for loop parameters. However, do not use a wide range of concentrations and a small incremental step unless you have a lot of time to waste!

5.4 User defined functions

In most dialects of BASIC it is possible for the user to define special functions. If this facility is available the user can define a function once, usually at the start of the program. Once it has been defined it can be used with any argument (a number, variable or arithmetic expression) in the remainder of the program. The general format of definition procedure is

DEF FNname (dummy argument) = expression

The DEF is an abbreviation of define. FN is short for function and name is the name of the function. In most dialects this must be a letter between A and Z. Thus the number of user defined functions is restricted to 26. The expression is the arithmetic expression which will be used to evaluate the function. The dummy argument can be a single variable or a list of variables. The actual variables used in the dummy argument is not important since when the function is evaluated the dummy variables will be replaced by other variables specified by the user. The user defined function is used in a similar way to other BASIC functions. It must not be used alone and requires an argument whose function it will evaluate. The use of user defined functions is best explained in an example.

Earlier in this chapter an equation was given for the calculation of the arcsine of a number. The expression could be used to define a function which could then be used whenever required. The calculation of arcsines can be very useful in certain statistical tests. Many of the parametric statistical tests require that certain conditions have been met, otherwise the results may be invalid. One of these assumptions is that the data used has been obtained from a population of measurements whose frequency distribution is normal. When the data used in such tests is in the form of percentages, which fall within the range 0–100%, then the normality assumption may be violated. Such data can be used if it is first transformed so that the resultant frequency distribution is normal. One method of transforming percentages is to use an arcsine transformation. The square root of each percentage is replaced by its arcsine. The figure calculated will be in radians. If this is converted to degrees then the transformed values will be in same order of magnitude (0–90) as the original 0–100. The use of a user defined function to transform percentages is shown below (note that not all of the program is shown).

```
10 REM DEMONSTRATION OF USER DEFINED FUNCTIONS
20 DEF FNA(X) = ( ATN ( X / SQR ( -X * X + 1 )) * 180 / 3.1416 )
    .
    .
100 REM  P(I) IS AN ARRAY OF PERCENTAGES, T(I) IS THE TRANSFORMED ARRAY
110 FOR I = 1 TO N
120 T(I) = FNA( P(I) / 100 )
130 NEXT I
```

Line 20 defines the function A, which is evaluated from the expression to the right of the = sign. The function is used in line 120 to transform an array of percentages, P(), to an array of their arcsine transformations, T(). The statement on line 120 should be read as: let the variable T(I) take the value of the function A for the argument P(I)/100. Note that although X was used in the definition of function A, it is not used on line 120. This is because X was used as a dummy variable and when the function is evaluated X is replaced by the new argument, which is P(I)/100 in this example. The function could be used again with a different variable or expression which would also take the place of the dummy argument X in the evaluation of function A. The percentages are divided by 100 to bring them into the valid range (0 < range < 1) for the function.

5.5 Summary

The following BASIC commands have been introduced in this chapter. (In the following examples X is the argument which is evaluated by the function. It may be replaced by any other legal variable or arithmetic expression which is within the valid numeric range for that function.)

SIN(X)   Evaluates the sine of an angle X, measured in radians.

COS(X)   As above, but evaluates the cosine.

TAN(X)   As above, but evaluates the tangent.

ATN(X)   As above, but evaluates the arctangent.

INT(X)   Evaluated as the greatest integer which is either less than or equal to the argument. It is usually restricted to a numeric range of −32767 to +32767.

SGN(X)   This function evaluates the sign of the argument and returns −1 if it is less than zero, 0 if the argument is zero and +1 if it is greater than zero.

61

ABS(X) This returns the absolute value of the argument. It is an unsigned number which is represented mathematically as $|x|$.

SQR(X) Evaluated as the square root of the argument and is therefore equivalent to X^0.5.

EXP(X) Evaluated as e raised to the power of the argument, i.e. e^x, e is normally stored as 2.718289 (correct to 6 decimal places).

LOG(X) This returns the natural logarithm of the argument, i.e. $\log_e x$.

RND(X) The syntax and use of this command differs between dialects but it is used to produce a pseudo-random number between certain limits.

DEF FN The user can define a function with this command. The syntax is:

DEF FN variable name (dummy variable) = expression

The dummy variable is replaced by the variable or variables contained with the function when it is used.

5.6 Problems

1. Write a function which can be used to evaluate the $\log_{10}$ of a number.

2. Write a program which can be used to determine the light intensity at specified depths in a body of water. This can be predicted from the following equation.

$$I_z = I_0 e^{-kz}$$

where I_z = light intensity at depth z (depths are in metres)
I_0 = light intensity at the surface
k = extinction coefficient

Typical values: surface light intensity 4000–20000 lux
k 0.04–0.05 for clear ocean; 0.1–0.2 for temperate oceans; 1.0 in coastal regions and 3–4.0 in turbid lakes.

3. The net assimilation of a crop can be estimated from:

$$E = \frac{W_2 - W_1}{T_2 - T_1} \times \frac{\log_e L_{A2} - \log_e L_{A1}}{L_{A2} - L_{A1}}$$

where E = net assimilation g m^{-2} unit time^{-1}
L_A = leaf area at times 1 and 2
T = times 1 and 2
W = plant weight at times 1 and 2 (g)

Write a program to estimate E when

$$T_2 - T_1 = 3 \text{ weeks}$$
$$W_1 = 240 \text{ g}$$
$$W_2 = 385 \text{ g}$$
$$L_{A1} = 1.9 \text{ m}^2$$
$$L_{A2} = 2.95 \text{ m}^2$$

4. The growth of many biological populations can be described by an equation of the following type.

$$y = \frac{a}{1 + b.e^{-kt}}$$

where y = number of organisms at time t; a, b and k would be determined from experimental data; e is the base of natural logarithms.

Write a program which can be used to provide a table of numbers of organisms against time for user entered values of a, b and k. Typical values would be in the following ranges: a, 200–300; b, 6–7; k, −0.1 to −0.2.

Chapter 6 Program Control Structures

6.1 Introduction

Until now all of the program examples have been constructed so that there was only one pathway through the instructions. Program execution started with the lowest numbered line and gradually progressed through the program by always executing the next highest line number. In fact, the computer has been used as little more than a glorified calculator. The power and flexibility of computers is a result of the fact that programs can be written in such a way that there are many pathways through the code. This is possible because there are program control structures which can be used to divert the sequence of program execution along different pathways. These program control structures fall into two distinct categories, unconditional and conditional commands.

6.2 Unconditional commands

There are two commands in this category, GOTO line number and GOSUB line number. When either of these commands is executed program control will be diverted to the line number specified in the command. This is why they are called unconditional control commands.

6.2.1 GOTO

The format of this command is:

    line number GOTO linenumber

for example,

```
100        GOTO 300
1030       GOTO 25
```

When this command is executed program execution is diverted to the specified line number, 300 and 25 respectively in the two examples. The program will then continue execution from this new position. It is, therefore, a very simple command to use. It is also a very easy command to abuse. The use of GOTOs is very much frowned upon unless alternatives are not available. A liberal use of GOTO within a program can produce a program structure which is very difficult to follow. The following example is a slight exaggeration of what can happen when the GOTO command is abused. Try to predict what the value of X will be when the program terminates!

```
10 LET X = 1
20 GOTO 100
30 LET X = X^2
40 GOTO 60
50 GOTO 80
60 LET X = 1
70 GOTO 50
80 LET X = 7
90 GOTO 120
100 LET X = 2
110 GOTO 30
120 PRINT "X = ";X
130 END
```

If the above program is run the final value of X would be 7. Imagine trying to follow what is happening to the values of several variables in a long program if it contains as many GOTO statements as above. (Such programs do exist!) The GOTO statement will, in future, only be used in those circumstances where there is no other command or program structure available.

6.2.2 GOSUB

This is similar to the GOTO command except that control is passed to a subroutine which is a block of instructions capable of performing one of three main functions. Firstly, subroutines can be used as the building blocks of programs. In previous examples the problems have been split into a number of blocks. In effect, each of these blocks were subroutines, although they were not identified as such within the programs. Subroutines can also be used to hold a block of instructions which is likely to be executed on several different occasions, thus avoiding the need to write repetitive code. The final use of subroutines is to improve program efficiency. This aspect will be covered in Chapter 14.

The format of the GOSUB command is:

line number   GOSUB linenumber

for example,

```
10   GOSUB 1000
120   GOSUB 500
1500   GOSUB 150
```

The beginning of a subroutine is not marked by a specific statement. It is, however, good practice to label the subroutine by placing a REM statement as the first instruction. The end of the subroutine is marked by a BASIC command, RETURN. (This has no connection with the RETURN key.) When a GOSUB command is executed the number of the program line holding the

GOSUB instruction is retained by the CPU. The subroutine is terminated when the RETURN instruction is executed. The program then continues execution from the line number immediately following that which issued the subroutine call. If a RETURN statement is executed when a subroutine call has not been issued an error message will be produced. This is because the CPU does not have a line number to return to. It is possible for subroutines to call other subroutines, a process which is called nesting. In such cases a RETURN command in a nested subroutine will return control to a line number in the subroutine which initially called it. Recursion is said to occur when a subroutine calls itself. This can be a very powerful facility but it is difficult to use in BASIC. The depth of recursion will depend upon how many subroutines can be nested by a particular interpreter.

BASIC subroutines are much less powerful than those available in Pascal and FORTRAN. They lack many of the features which could make them more useful. Some of these deficiencies are discussed in Chapter 10.

Example 6.1

In this example the simple statistics program described in Chapter 4 (Example 4.5) will be converted to use subroutines.

```
10 REM               SIMPLE STATISTICS PROGRAM, USING SUBROUTINES
11 REM     VARIABLES D = STANDARD DEVIATION, M = MEAN, N = SAMPLE SIZE
12 REM               S = SUM OF SQUARES, X(I) = ARRAY OF VALUES
13 REM               V = VARIANCE
14 REM
19 REM               INITIALISATION
20 GOSUB 100
29 REM               USER INFORMATION
30 GOSUB 200
39 REM               DATA ENTRY
40 GOSUB 300
49 REM               CALCULATIONS
50 GOSUB 400
59 REM               RESULTS OUTPUT
60 GOSUB 500
70 END
100 REM -----------INITIALISATION---------------------------------
110 LET D = 0
120 LET M = 0
130 LET N = 0
140 LET S = 0
150 LET V = 0
160 DIM X(50)
170 RETURN
200 REM -----------USER INFORMATION-------------------------------
210 PRINT "CALCULATION OF SIMPLE STATISTICS"
220 PRINT
230 PRINT "THIS PROGRAM WILL CALCULATE THE MEAN, STANDARD DEVIATION AND"
240 PRINT "VARIANCE FOR A SAMPLE OF UPTO 50 NUMBERS."
250 PRINT
```

```
260 RETURN
300 REM -----------DATA ENTRY-----------------------------------
310 PRINT "HOW LARGE IS THE SAMPLE";
320 INPUT N
330 PRINT "ENTER " ; N ; " NUMBERS"
340 FOR I = 1 TO N
350 PRINT I;"   ";
360 INPUT X(I)
370 NEXT I
380 RETURN
400 REM ------------CALCULATIONS----------------------------------
410 FOR I = 1 TO N
420 LET T = T + X(I)
430 LET T2 = T2 + X(I)^2
440 NEXT I
450 LET M = T / N
460 LET S = T2 - T^2 / N
470 LET V = S / (N - 1)
480 LET D = SQR(V)
490 RETURN
500 REM ----------RESULTS OUTPUT------------------------------------
510 PRINT
520 PRINT "SAMPLE SIZE        = " ; N
530 PRINT "MEAN               = " ; M
540 PRINT "VARIANCE           = " ; V
550 PRINT "STANDARD DEVIATION = " ; D
560 RETURN
```

The main program occupies lines 10 to 70. When a program is set out in this way it is relatively easy to visualise the sequence of operations. Note that the END of the program is before the first subroutine. If it was not in this position the first subroutine (lines 100 to 170) would be executed twice. The programming instructions are identical to those in Example 4.5 except for line 480. The square root function, SQR, has replaced $\wedge 0.5$ in the earlier program.

When the program is run the first GOSUB instruction, on line 20, will transfer execution to line 100. The following lines will be executed ending at line 170 with the RETURN command. This transfers execution to line 29, which is the next line following line 20 (from where the subroutine was called). The GOSUB command on line 30 transfers control to line 200, the second subroutine. This subroutine is terminated by the RETURN command on line 260. The next line to be executed would be line 39. This is followed by the third GOSUB on line 40 and so on.

6.3 Conditional commands

There are several conditional commands available in most BASIC dialects. A conditional command is one in which the action taken depends upon the results of a test. This discussion will start with the two simplest. These are ON ... GOTO and ON ... GOSUB.

6.3.1 ON ... GOTO

The syntax of this command is:

line number ON variable GOTO 1st line number, 2nd line number, ...nth line number

for example,

```
100 ON X GOTO 200, 300, 400, 500
3050 ON T1 GOTO 100, 200, 1000, 10000, 50
```

In the first example X would have a value between 1 and 4. If X was 1 the transfer would be to line 200. If X was 2 transfer would be to line 300, etc. In the second example T1 would have a value of between 1 and 5. The following transfers would take place.

Value of T1	Control transferred to line
1	100
2	200
3	1000
4	10000
5	50

6.3.2 ON ... GOSUB

This is a very similar command to ON ... GOTO, the difference being that control is transferred to a subroutine. The syntax of the command is:

line number ON variable GOSUB 1st line number, 2nd line number, ...nth line number

for example,

```
25 ON W GOSUB 1000, 300, 6000, 250, 300
```

In this example W would have a value between 1 and 5. Note that if W is equal to 2 or 5 the subroutine starting on line number 300 is called.

Example 6.2

This program uses the ON ... GOSUB command to call one of four subroutines depending upon a value entered by the user. This is not a complete program but it could form part of a comprehensive statistics program. It uses a menu system to present information to the user, a menu being a list of options from which the user selects one option. Menus are frequently used to make programs 'user friendly', i.e. they simplify the exchange of information between the user and the program.

It is often desirable to transform values before the computation of some statistics. Transformation involves carrying out a mathematical treatment on all of the original data. This is usually necessary if the frequency distribution of the original data is not consistent with that required by the underlying assumptions of the statistical test. This is not data fixing! In the following example each transformation is contained within a separate subroutine and converts a range of frequency distributions into approximations of the normal distribution.

Four examples of the original and transformed frequency distributions are shown in Fig. 6.1. In this program the data to be transformed is already in the program. It is stored as an array of X values which has N elements.

```
100 REM ----------------TRANSFORMATION MENU-------------------------
110 REM                  CLEAR THE SCREEN
120 PRINT"SELECT ONE OF THE FOLLOWING TRANSFORMATION OPTIONS"
130 PRINT
140 PRINT"    1. SQUARE ROOT"
150 PRINT"    2. NATURAL LOGARITHM"
160 PRINT"    3. ARCSINE"
170 PRINT"    4. LOG( X / 1 - X )"
180 PRINT"    5. NONE"
190 PRINT
200 PRINT"ENTER THE NUMBER REQUIRED ";
210 INPUT O
220 ON O GOSUB 1000, 2000, 3000, 4000, 300
300 REM ------------------MAIN CALCULATIONS--------------------------
  .
  .
  .
999 END
1000 REM ----------------SQUARE ROOT TRANSFORMATION--------------------
1010 FOR I = 1 TO N
1020 LET X(I) = SQR ( X(I) )
1030 NEXT I
1040 RETURN
2000 REM ----------------NATURAL LOG TRANSFORMATION--------------------
2010 FOR I = 1 TO N
2020 LET X(I) = LOG ( X(I) )
2030 NEXT I
2040 RETURN
3000 REM ----------------ARCSINE TRANSFORMATION-----------------------
3010 FOR I = 1 TO N
3020 LET X(I) = ATN (X(I) / SQR ( - X(I) * X(I) + 1 ) )
3030 NEXT I
3040 RETURN
4000 REM ----------------LOG ( X / 1 - X ) TRANSFORMATION--------------
4010 FOR I = 1 TO N
4020 LET X(I) = LOG ( X(I) / ( 1 - X(I) ) )
4030 NEXT I
4040 RETURN
```

A value for the variable O, between 1 and 5, is entered on line 210. This is then used by the ON ... GOSUB command on line 220 to determine which of the transformation subroutines is required. If 5, no transformation, is selected control passes to the main calculations which start on line 300. The original distributions   transformed distributions

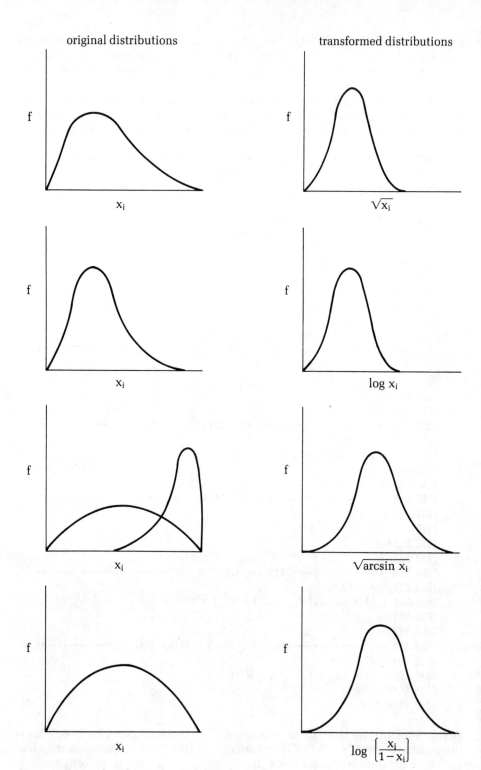

Fig. 6.1 Original and transformed distributions

RETURN commands, which terminate each subroutine will also transfer control to this line since it is the line which follows line 220 (from where the subroutines are called).

Sample run

```
>RUN

SELECT ONE OF THE FOLLOWING TRANSFORMATION OPTIONS
    1. SQUARE ROOT
    2. NATURAL LOGARITHM
    3. ARCSINE
    4. LOG( X / 1 - X )
    5. NONE

ENTER THE NUMBER REQUIRED ?1
```

The remainder of the program would now follow. The calculations would be carried out on an array of X values which are now equal to the square root of their initial values.

Note

The arcsine and the log(x/1−x) transformations will only work if X(I) has a value of between 0 and +1. In most cases it would, therefore, be necessary to carry out some other manipulations on the data to bring the values within this range.

6.3.3 IF ... THEN

All CPUs have the ability to compare two numbers. Essentially this involves a subtraction. If the result is negative the second number is larger. If the result is zero the numbers must be equal. Finally a positive result indicates that the first number is larger. This relatively simple operation can produce some very powerful commands. One of these is the IF ... THEN statement. The format of this statement is:

line number IF a condition is true THEN take the following action

The computer determines if the condition is true by using one of the BASIC relational operators which specify a relationship between two numbers. The list of available operators is shown in Table 6.1.

Table 6.1 BASIC relational operators

Relationship	Mathematical symbol	BASIC relational operator
equal to	$=$	$=$
less than	$<$	$<$
less than or equal to	$\leq$	$< =$ or $= <$
greater than	$>$	$>$
greater than or equal to	$\geq$	$> =$ or $= >$
not equal to	$\neq$	$< >$

It is, therefore, possible to test conditions of the following type.

$$X > Y ? \qquad T1 = T2 ? \qquad N < 5 ?$$

These conditions are tested by using the CPU's ability to compare numbers. In the first example, if X is greater than Y then $X - Y$ will yield a positive result. A result which is zero or negative would indicate that the condition is false, i.e. X is not greater than Y. The computer uses a flag (or marker) to indicate the validity of a condition. A flag is a memory location or variable which is set to a particular value depending upon the outcome of a test condition. The flag values used for true and false conditions vary between computers. They are often 1 and 0 or -1 and 0.

If the condition is true the operation which follows the THEN part of the statement will be executed. If the condition is false this operation will not be executed and control will pass to the next program line. The IF ... THEN construction can be used very effectively to control a repetitive process of indeterminate length. The following example can be used to enter any number of values into a one dimensional array (as long as sufficient space has been reserved with an appropriate DIM statement).

```
100 REM        DEMONSTRATION OF THE IF THEN STATEMENT
101 REM        T = 0 AND C = 1. INITIALISED AT THE START OF THE PROGRAM
110 PRINT "ENTER THE VALUES, INDICATE THE END OF THE LIST BY ENTERING -1"
120 INPUT T
130 IF T = -1 THEN GOTO 170
140 LET X(C) = T
150 LET C = C + 1
160 GOTO 120
170 REM        THE REMAINING PART OF THE PROGRAM STARTS HERE
```

The relevant part of this program occupies lines 120 to 160. A value is entered on line 120 and stored as a temporary variable, T. This value is then compared with a flag value of -1 by the IF part of the IF ... THEN statement. If the condition is true, i.e. T equals -1, the GOTO command is executed and control passes to line 170 and the rest of the program. If the condition is false, i.e. T does not equal -1, the GOTO command is ignored and lines 140 to 160 are executed. The value is transferred from its temporary variable, T, to its permanent variable, X(C), by line 140. A counter, C, is incremented by one on line 150. This keeps track of how many values have been entered and provides the array subscripts. There is an unconditional GOTO command on line 160 that returns control to line 120.

6.3.4 IF ... THEN ... ELSE

This is a useful extension of the IF ... THEN construction. Unfortunately it is not universally available. The format of this statement is:

line number IF test condition THEN do this if true ELSE do this if false

for example,

```
100 IF X > 0 THEN PRINT "POSITIVE" ELSE PRINT "NOT POSITIVE"
720 IF N > 30 THEN N1 = N ELSE N1 = N - 1
5000 IF C < > 1 THEN GOSUB 6000 ELSE GOSUB 7000
```

In the first of these examples either POSITIVE or NOT POSITIVE would be printed depending upon whether X was greater or less than zero. The second example compares N with 30 and then assigns one of two values to a second variable N1. If X was greater than 30 then N1 would be assigned the current value of N otherwise N1 would take the value of N−1. In the final example the subroutine starting at line 7000 would be called only if C equalled 1. (This could also have been written as IF C = 1 THEN GOSUB 7000 ELSE GOSUB 6000.) If the ELSE option is unavailable a GOTO command, or another IF ... THEN construction, would be required to perform the same operations. For example,

```
100 IF X > 0 THEN PRINT "POSITIVE"
105 IF X < 0 THEN PRINT "NEGATIVE"
```

6.3.5 Logical operators

Logical operators are expressions which can be used to combine test conditions into more complex constructions. There are usually a maximum of four expressions available in BASIC but only two are commonly used. The logical operators are:

NOT
AND
OR
XOR (eXclusive OR)

The two most frequently used logical operators are AND and OR. They have a meaning which is very similar to normal English. When two expressions are linked by a logical operator it is the combination of expressions which is assessed as true or false.

Example 6.3

```
IF X>0 AND Y=5 THEN PRINT "TRUE" ELSE PRINT "FALSE"
X = 3 and Y = 5              TRUE
X = 0 and Y = 5                           FALSE
X = 6 and Y = 7                           FALSE
```

When AND is used both of the linked expressions must be true for the combination to be true. If OR is used then one or both of the linked expressions must be true for the combination to be true. The XOR statement differs from the OR statement in that the combined expression is true only if one of the linked expressions is true. The NOT statement is used with a single expression and can be slightly confusing. An expression of the type NOT A = B would be true if A did not equal B. If A was equal to B then the expression NOT A = B would be false.

These relationships can be summarised in a truth table as shown in Table 6.2

Table 6.2 Logical operators truth table

Operator	Expression 1	Expression 2	Result
NOT	True		False
	False		True
AND	True	True	True
	False	True	False
	True	False	False
	False	False	False
OR	True	True	True
	False	True	True
	True	False	True
	False	False	False
XOR	True	True	False
	True	False	True
	False	True	True
	False	False	False

Logical operators can be used in conjunction with conditional control statements to provide error checking routines. Such a routine is a section of code which is used to monitor inputs and verify that they fall within a valid range. Example 6.2 requested the user to enter the number of the transformation required. Additional lines can be added between lines 210 and 200 to check that the number entered is between 1 and 5.

```
211 IF O > 0 AND O < 6 THEN GOTO 220
212 PRINT "PLEASE RESELECT A NUMBER BETWEEN 1 AND 5"
213 GOTO 210
```

If a number between 1 and 5 has been assigned to the variable O the combined expression on line 211 would be true. However, if the number was less than 0 or greater than 5 it would result in this expression being evaluated as false. (Note that a very common fault when writing a line such as 211 is to miswrite it as IF O > 0 AND < 6 . This cannot be evaluated and would result in a program crash.)

6.3.6 Precedence

Since the logical and relational operators are included in expressions which also contain algebraic operators, it is important to understand the order of precedence for these operators. The complete sequence of precedence is shown in Table 6.3.

Table 6.3 Order of precedence of operators

Order of precedence	Operator
1	( )
2	+ − [unary operators]
3	^
4	× /
5	+ −
6	> < >= => <= =< <> >< =
7	AND
8	OR

Note that operators with the same order of precedence would be evaluated on a left to right basis. Unary operators are used to assign positive or negative values to variables. Since unsigned numbers are positive by default the unary + operator is seldom used.

6.4 Summary

The following BASIC commands have been introduced in this chapter.

GOTO   This is an unconditional control command which forces a branch to a specific line number, e.g. GOTO 1200, GOTO 534, GOTO 15.

GOSUB   GOSUB is a similar command to GOTO in that it forces an unconditional branch to a specified line number. This line number marks the beginning of a subroutine. A subroutine is a block of code which can be used to give a program structure. It may also be used to hold a block of instructions which will be executed several times. See RETURN.

RETURN   RETURN is the statement which marks the end of a subroutine. When it is executed it will force a branch to the line following that which called the subroutine. It does not, therefore, always return control to the same point in a program.

ON ... GOTO   This is the first of the conditional statements. It forces a branch to a line number depending upon the value of a variable, e.g. ON J GOTO 100, 200, 250, 1000, 100. If the variable is equal to 1 the branch will be to the first line number, if it is 2 the branch will be to the second line number, etc.

ON ... GOSUB   Very similar to ON ... GOTO except that control is passed to a line number which marks the start of a subroutine.

RELATIONAL   These operators are used to test the relationships
AND LOGICAL   between variables and expressions.
OPERATORS

                       Relational operators
                           $=$ equal
                           $< >$ not equal
                           $<$ less than
                           $>$ greater than
                           $< =$ less than or equal
                           $> =$ greater than or equal

Logical operators
NOT logical 'not'
AND logical 'and'
OR logical 'or'
XOR logical 'exclusive or'

IF ... THEN
There are two parts to this command. The first part, IF, establishes a condition which will be tested. If the condition is false, control will immediately pass to the next line. However, if the condition is true, the operation which follows the THEN part of the command will be executed. Some examples are shown below.

```
IF X > Y THEN Z = X + 2
IF T1 < 30 THEN PRINT " LESS THAN THIRTY"
IF D - 5 < > F5 / 360 THEN GOTO 300
```

IF ... THEN ... ELSE
This is an extension of the IF ... THEN command. An extra statement, ELSE, terminates the command. When the condition specified by the IF part of the command is false, the statement which follows ELSE will be executed. For example:

```
IF B = 0 THEN H = H^2 ELSE H = B
IF K > 7 THEN GOTO 450 ELSE END
IF INT ( A / 10 ) < > 5 THEN PRINT "YES" ELSE
PRINT "NO"
```

6.5 Problems

1. Write a subroutine which can be used to search through an array which contains systolic blood pressure records for a number of patients. The subroutine should print out the patient numbers (array subscripts) of those with a pressure greater than 125 mm Hg.

2. Modify the subroutine from Problem 1 to include the following. Print out the patient numbers if the systolic blood pressure is greater than 115 mm Hg and the body weight is greater than 85 kg. (Hint, use two arrays.)

3. Write the following subroutines. Assume that data has already been entered into any arrays that are required.
   i) A menu which offers the following options:
      a) select smokers
      b) select heavy smokers (number of cigarettes per day > 20)
      c) select heavy smokers who do not drink alcohol
      d) select individuals who are male, heavy smokers, and drinkers.
   ii) The subroutines which would be called from the menu.

Chapter 7 Strings

7.1 Introduction

Until now we have considered computers to be machines which are capable of manipulating numbers. Computers can also perform various operations on characters such as symbols and letters of the alphabet. Lists of these characters are known as character strings or more simply, strings. All of the alphanumeric (letters and numbers) characters and symbols which a computer can recognise can be represented by a numeric code system. The most common type of code is the ASCII (American Standard Code for Information Interchange). All of the upper- and lowercase letters, the numbers 0 to 9, numeric operators and punctuation marks, plus a number of special control codes are represented by the numbers 0 to 127 (bits 0 to 6 of an eight-bit byte). The full ASCII code is shown in Appendix B. Consequently, when a computer stores information about strings it does so as a sequence of numbers which correspond to the ASCII codes of the characters in the string. The computer treats these numbers in a different way than the numeric variables which have been considered in the earlier chapters. Even though all characters are stored internally as a number it is impossible to carry out arithmetic operations on them.

This chapter contains three main sections. The first introduces some of the more common string handling commands. The second section considers how strings may be formatted to a required specification. The final section consists of several short programs which illustrate some of the uses of strings and string commands in biologically orientated programs.

7.2 BASIC string commands

7.2.1 String variables

A string variable can generally contain between 0 and 255 characters (there are exceptions to this). A string variable is distinguished from a numeric variable by including a $ or £ sign as the last character in the variable name. All other rules concerning the nature of variable names are the same for numeric and string variables. Thus the variable A$ would be a string variable which is independent of the numeric variable A. Characters are assigned to string variables in the same way as values are assigned to numeric variables. Thus the following are valid statements.

BASIC statement	Explanation
INPUT Q$	input a string to be called Q$
READ T1$	read in a string to the variable T1$
LET B9$ = "string"	let the variable B9$ be assigned the string of characters contained within the double quotes

Note that when a string is assigned by a LET statement the string must be enclosed within double quotes. If the quotes were missing the computer would respond with a message such as TYPE MISMATCH. In other words an attempt has been made to assign something that the computer did not recognise as a string to a string variable. If a READ statement is used the string or strings will be held on a DATA line. The quotes are not always essential for strings held on DATA lines.

Since the numbers 0 to 9 can be represented by their ASCII codes (48–57) they can be used in strings. The distinction between a numeric variable and a string variable composed of numbers is very important. The computer treats them in completely different ways. For example, if A = 3 and B = 4 the sum of A + B would be 7; but if A$ = "3" and B$ = "4" then A$ + B$ would equal "34". This is because the + symbol does not force an addition when used with strings. Instead it produces a concatenation of the strings, in other words the second string is appended to the end of the first string. Thus the A$ + B$ operation appended B$ onto the end of A$ to produce the result "34". Prove it to yourself by trying out "4" + "3"!

Some examples of string variables and concatenations are shown in Table 7.1.

Table 7.1 String variables and concatenations

A$ = "HOMO"	B$ = "SAPIENS"	A$ + B$ = "HOMOSAPIENS"
A$ = "HOMO"	C$ = " SAPIENS"	A$ + C$ = "HOMO SAPIENS"
T1$ = "D" T2$ = "N" T3$ = "A"		T1$ + T2$ + T3$ = "DNA"
P$ = "12345"	W$ = "6789"	P$ + W$ = "123456789"
D$ = "760"	U$ = " MM HG"	D$ + U$ = "760 MM HG"
N$ = "89.5"	P$ = " %"	N$ + P$ = "89.5 %"

Note that spaces can be included within strings and may be essential to get a meaningful result from a concatenation. It is also possible to mix numbers, symbols and letters within a string. This is because in this format numbers are treated as characters which have no numeric value.

String arrays can usually be created by using the DIM statement. Some dialects of BASIC do not, however, allow arrays to be used for string storage. The method of dimensioning a string array is identical to that used with numeric arrays except that the array name will be terminated by a $ symbol. The following would all be valid statements: DIM A$(30), DIM X1$(6), DIM B$(100).

7.2.2 ASC and CHR$

These two commands are related in that they have opposing functions.

ASC(string variable) returns the ASCII code of the first character in the string variable. Thus if A$ = "9", ASC(A$) would equal 57 which is the ASCII code for "9". Because only the first character is used in the evaluation ASC("976") would also be 57.

CHR$(number between 0 and 255) returns the character whose ASCII code is the numeric integer expression within the parentheses. Thus CHR$(57) would return "9", CHR$(75) would be "K" and CHR$(107) would return "k". The number must be in the range 0 to 255, these being the limits for the ASCII code. The first 127 codes (0 to 127) are usually consistent between various models of computer but the remaining values are used differently by each manufacturer. Consult the computer handbook to determine which characters are represented by the codes 128 to 255.

The CHR$ command can be a useful way of including control codes within a program. For example, if the statement PRINT CHR$(7) was executed on an APPLE II microcomputer nothing would be printed but a 'bell' would ring! Similarly the statement PRINT CHR$(8) would result in the screen cursor back-spacing one position.

7.2.3 VAL and STR$

These two commands are also opposites. VAL converts a string of numeric characters into their numeric value while STR$ converts a numeric value into a string. Some examples are shown below.

LET X = VAL(X$)	If X$ = "234.5" X would have a value of 234.5.
LET T = VAL(A$) + VAL(B$)	If A$ = "3" and B$ = "4" T would have a value of 7.
LET T = VAL(A$ + B$)	T would now have a value of 34 because the strings A$ and B$ would be concatenated before the evaluation was carried out.
LET T$ = STR$(432.8)	T$ would become the character string "432.8".
LET N$ = STR$(SQR(N))	If N was 25 N$ would be the string "5".

7.2.4 LEN

This function evaluates the length (number of characters) of a string. LEN(A$) would return the number of characters in the string A$. Thus if A$ = "1,000,345", LEN(A$) would return a value of 9 since there are 9 characters in the string (7 numbers plus 2 commas). Note that the string " 1,000,345 " has a length of 11 characters since there are two additional spaces in the string which would be counted as characters.

7.2.5 LEFT$ and RIGHT$

As their names suggest these two commands are associated with the left and right hand parts of a string. Both of these commands require two arguments to operate on. They require a string name and the number of characters to be considered.

The statement PRINT LEFT$(A$,3) would print the first three characters, counting from the left, of the string A$. PRINT RIGHT$(A$,3) would print the first three characters, counting from the right, of the string A$. They can be used to break down strings or produce new strings from parts of existing strings. Consider the following short program.

```
10 REM DEMONSTRATION OF LEFT$ COMMAND
20 LET A$ = "ASTERIAS"
30 FOR I = 1 TO LEN(A$)
40 PRINT LEFT$(A$,I)
50 NEXT I
60 END
```

If this program is run the following output would be produced.

```
>RUN

A
AS
AST
ASTE
ASTER
ASTERI
ASTERIA
ASTERIAS

>OK
```

Similar operations can be carried out using the RIGHT$ command. Note how the LEN command was used on line 30 to set the end value for the loop. On each execution of the loop I was incremented by 1 and, therefore, one more character from the string A$ was printed.

7.2.6 MID$

This is a similar command to the previous two except that it is used to pick out parts of a string starting from any position within the string. This command functions, therefore, as MIDdle$. Since it can be used to pick a substring starting from any position, both the length of the substring and starting position must be specified. This command has, therefore, three arguments: the string name, the start of the substring and the length of the substring.

The previous program can be amended to use the MID$ command. If line 40 is altered to read:

```
40 PRINT MID$(A$,I,3)
```

and the program is run the following output would be produced.

```
>RUN

AST
STE
TER
ERI
RIA
IAS
AS
S

>OK
```

The string A$ has been dissected so that the three characters to the right of position I are printed. This can be a very useful operation since it can be used to search through a long string for any specified substring. A biological application of this operation is given in section 7.4.

7.2.7 GET and INKEY$

These two commands perform essentially the same operation but they are not available in all dialects. They are used to input a character from the keyboard without the need for a subsequent carriage return. They are not strictly equivalent because the GET command can be used with numeric and string variables while the INKEY$ command works only with string variables.

Normally, a character entered from a keyboard is stored in a buffer and only transferred to the program when the RETURN key, or its equivalent, is pressed. These commands circumvent this procedure. When either of these commands is executed the CPU scans the keyboard to see if any key is being pressed. If a key is depressed as the INKEY$ statement is executed its value will be transferred to the program. If no key was pressed the next line of the program will be executed (normally). Since these commands are executed very rapidly it is usually necessary to put in a delay loop so that the program waits until a suitable key has been depressed. An example application is shown below.

```
       100 REM   A SCREENFUL OF INFORMATION HAS BEEN PRESENTED TO THE USER
       110 REM   THERE IS MORE TO COME AND THIS WILL BE PRINTED WHEN THE
       120 REM   USER HAS READ THE CURRENT SCREEN CONTENTS.
       130 REM   LINE 160 PROMPTS THE USER TO PRESS ANY KEY WHEN READY TO
       140 REM   CONTINUE. LINES 170 & 180 MONITOR THE KEYBOARD WAITNG FOR
       150 REM   A SIGNAL FROM THE USER.
       160 PRINT "PRESS ANY KEY TO CONTINUE"
       170 A$ = INKEY$
   or  170 GET A$
       180 IF A$ = "" THEN GOTO 170
       190 REM   A KEY HAS BEEN PRESSED THEREFORE PRINT THE NEXT SCREENFUL.
```

Line 170 scans the keyboard and assigns a character to the variable if a key was pressed as this line was executed. (Both command formats are given.) Line 180 compares the string A$ with a null string. If no key has been pressed A$ will be empty and hence the condition tested by the IF statement will be true. Consequently, control will return to line 170. The program will continue to loop between these two lines until any key is pressed. At this point A$ will no longer be null and the IF condition will be false. This allows line 190 to be executed. Note that even if a control key such as the RETURN key is pressed A$ will no longer be null since all the control keys have ASCII codes which would be stored in A$. An application of these commands is given in section 7.4. Users of the BBC microcomputer should consult their manual since the GET and INKEY functions have a different implementation.

7.2.8 String logical and relational operators

All of the logical and relational operators used with numeric values can also be used with strings. The complete list of operators is given in Tables 6.1 and 6.2.

If two strings are equal there must be a match between each of their constituent characters. Equality is verified by comparing the ASCII codes of all characters, including spaces, in both strings. It is also possible to determine if one string is greater or less than another string. The strings do not need to be numeric for this operation to be valid. The first characters in both strings are compared. The string which begins with a character having the lowest ASCII code is said to be the lesser of the two. If the first characters are identical the comparison moves to second characters; if these are identical the third characters are compared. This continues until a difference is located. The ASCII code is constructed so that the code for 0 is less than that for 1, etc. Similarly the code for A is less than that of B, etc. The relational operators can, therefore, be used to sort a list into alphabetical order by carrying out a series of comparisons. A numeric sorting program is given in Chapter 10. This can be easily adapted to sort strings into alphabetical order. It should be noted that because of the way the code is constructed all uppercase letters are less (have lower ASCII codes) than the lowercase letters. If a string sort is likely to come into contact with letters of both case it will be necessary to include a small detection and amendment routine.

7.3 Print formatting

When a program produces an output comprising many separate numbers the display can become very cluttered and difficult to interpret. This can be avoided if the numbers are formatted before printing. The easiest method of producing ordered tables is unfortunately not a universal component of BASIC dialects. This is the PRINT USING command. Inevitably this command varies somewhat in its implementation and use.

7.3.1 PRINT USING

The PRINT USING command is aimed mainly at business users since it allows cheques and other financial documents to be printed in specific formats. These aspects of this command will not be discussed here. The command can also be used as an aid to the orderly tabulation of numbers.

The PRINT USING command is used to establish a printing format for numbers. This means specifying the number of characters to be printed (including the decimal point) and how many digits there are to be on either side of the decimal point. If a number has fewer digits than the specified format it will be right justified before printing. Right justification is the insertion of spaces to the left hand side of the number so that a column of numbers will become vertically aligned. The syntax of this command is:

PRINT USING "format"; list of numbers

The format is specified by using the # symbol to represent digits. Note that these must be included in double quotes since the computer treats these as a string variable. The effects of the PRINT USING command are demonstrated in the next short program which reads in a list of numbers and then prints them.

```
10 REM    DEMONSTRATION OF THE PRINT USING COMMAND
20 FOR I = 1 TO 10
30 READ X(I)
40 REM    PRINT COMMAND GOES HERE, SEE TABLE 7.2 FOR DIFFERENT FORMATS
50 NEXT I
60 DATA 123.45,3.46,56.737,1.1,100,2367.5423,66.723,9.999,5,201.451
70 END
```

Table 7.2 Different formats of the PRINT USING command

4 PRINT X(I)	40 PRINT USING "###.##";X(I)	40 PRINT USING "####.#";X(I)
123.45	123.45	123.5
3.46	3.46	3.5
56.737	56.74	56.7
1.1	1.10	1.1
100	100.00	100.0
2367.5423	%2367.54	2367.5
66.723	66.72	66.7
9.999	10.00	10.0
5	5.00	5.0
201.451	201.45	201.5

There are several points of note in Table 7.2. The PRINT USING commands, by forcing the decimal points to align, have produced lists which are much easier to read. The first PRINT USING command specifies that there should be three digits to the left of the decimal point and two to the right. Numbers smaller than 100 are, therefore, preceded by the appropriate number of spaces. If the number of digits to the right of the decimal point is less than two, the missing digits are replaced by zeros. If the number of digits to the right of the decimal point is greater than that specified, the number is rounded as appropriate. Note that unlike the INT function this produces a true rounding. If a number is larger (numerically, not the number of digits) than the field allows, a % sign is printed in front of the number. Rounding can produce some unexpected results. The value 9.999 was rounded off to 10 in both PRINT USING examples. If a value of 999.999 had been included this would have been rounded off to 1000.00 by the first PRINT USING example. Since this is now too large for the print field it would have been printed as %1000.00.

7.4 Example programs

7.4.1 Student proof data entry

If a group of people is entering numbers into a program it is almost certain that one of them will press the wrong key. This will often cause the program to crash. It is advisable, therefore, to include in all programs error checking routines that will reduce the risk of user errors. The following program uses the INKEY$ function to enter numbers into a program. The program is written such that only 11 characters are recognised and accepted by the program. These are the numbers 0 to 9 and the RETURN key.

The algorithm is quite simple. The INKEY$ function is used to scan the keyboard. If a key is pressed the ASCII code of the character is checked to determine if it is valid. Valid ASCII codes are 48 to 57 (the numbers 0 to 9) and 13 (the RETURN key). If a key press inputs a character outside these values it will be ignored. If the RETURN key has not been pressed and the ASCII code is valid, the character is concatenated onto a string which is gradually assembled. When the RETURN key is pressed the string is complete and can be converted into a number.

```
1000 REM    IDIOT PROOF DATA ENTRY - INTEGER NUMBERS ONLY
1001 REM    T$ - TEMPORARY STRING USED WITH INKEY$
1002 REM    N$ - THE STRING INTO WHICH NUMBERS ARE CONCATENATED
1010 LET N$ = ""
1011 LET T$ = ""
1020 T$ = INKEY$
1021 IF T$ = "" THEN GOTO 1020
1030 IF ASC(T$) = 13 THEN GOTO 1070
1031 IF ASC(T$) < 48 OR ASC(T$) > 57 THEN GOTO 1020
1040 LET N$ = N$ + T$
1050 PRINT T$;
1060 GOTO 1020
1070 LET N = VAL(N$)
1080 The rest of the program starts here
```

The two string variables are initialised to null strings on lines 1010 and 1011. The keyboard scanning routine occupies lines 1020 and 1021. If a key is pressed its ASCII code is evaluated and compared with that of the RETURN key on 1030. If the RETURN key was not pressed line 1031 is executed. This line determines if the key was invalid by checking if the ASCII code is less than 48 or greater than 57. If the key was invalid T$ is not used and control returns to the keyboard scan. If T$ was a number it is concatenated onto the end of the number string N$. Since INKEY$ does not print the character onto the screen this is done by line 1050. The semicolon suppresses the carriage return which normally follows the PRINT command and hence the string N$ gradually appears on the screen. If the RETURN key was pressed the string is complete and control transfers to line 1070 where it is evaluated and saved as the numeric variable N.

7.4.2 Analysis of amino acid sequences

The string commands can be used in programs which investigate various properties of protein amino acid sequences. These sequences can be represented in various ways but the simplest is to use the one letter code. Table 7.3 lists the 20 amino acids and the single and three letter codes which are recognised internationally. Amino acid sequences can be found in various publications and it would be a relatively simple task to build up a library of such sequences for future investigations. Similar operations can be carried out on DNA base sequences (see Chapter 12 for more details).

The first program asks the user to input a sequence of amino acids (as single letters). The program then searches through the amino acid sequence of a particular protein and lists the points at which the requested sequence occurs.

Table 7.3 Amino acids and their codes

Amino acid	3 letter code	1 letter code	Amino acid	3 letter code	1 letter code
Alanine	Ala	A	Isoleucine	Ile	I
Arginine	Arg	R	Leucine	Leu	L
Asparagine	Asn	N	Lysine	Lys	K
Aspartic acid	Asp	D	Methionine	Met	M
Asn and/or Asp	Asx	B	Phenylalanine	Phe	F
Cysteine	Cys	C	Proline	Pro	P
Glutamine	Gln	Q	Serine	Ser	S
Glutamic acid	Glu	E	Threonine	Thr	T
Gln and/or Glu	Glx	Z	Tryptophan	Trp	W
Glycine	Gly	G	Tyrosine	Tyr	Y
Histidine	His	H	Valine	Val	V

```
10 REM     AMINO ACID SEQUENCE IDENTIFICATION PROGRAM
11 REM     VARIABLES USED
12 REM     N$ - NAME OF PROTEIN              S$ - AMINO ACID SEQUENCE
13 REM     A$ - STRING TO BE SEARCHED FOR    Q$ - USED WITH INKEY$
20 REM ---------- INITIALISATION -----------------------------------------
21 LET N$ = ""
22 LET S$ = ""
23 LET A$ = ""
24 LET Q$ = ""
30 REM ---------- READ IN NAME OF PROTEIN AND ITS AMINO ACID SEQUENCE ---
32 READ N$
34 IF N$ = "END" THEN END
36 READ S$
40 REM ---------- USER ENTERS THE STRING TO BE SEARCHED FOR -------------
41 PRINT "ENTER THE AMINO ACID REQUIRED AS SINGLE LETTERS WITH NO SPACES";
42 INPUT A$
50 REM ---------- SEARCH THE SEQUENCE FOR THE OCCURRENCE OF A$ ----------
52 PRINT "STRING "; A$; " OCCURS AT THE FOLLOWING POSTIONS IN "; N$
54 FOR I = 1 TO LEN( S$ )
56 IF MID$(S$, I, LEN( A$ ) ) = A$ THEN PRINT I
58 NEXT I
60 REM ---------- ANOTHER GO ? --------------------------------------------
61 PRINT
62 PRINT "PRESS E TO END THE PROGRAM, ANY OTHER KEY TO CONTINUE"
63 Q$ = INKEY$
64 IF Q$ = "" THEN GOTO 63
65 IF Q$ = "E" THEN END
66 PRINT "PRESS P FOR ANOTHER PROTEIN SEQUENCE OR S TO TEST ANOTHER AMINO"
67 PRINT "ACID SEQUENCE AGAINST ";N$
68 Q$ = INKEY$
69 IF Q$ = "" THEN GOTO 68
70 IF Q$ < > "P" AND Q$ < > "S" THEN GOTO 68
71 IF Q$ = "P" THEN GOTO 32 ELSE GOTO 40
999 END
1000 DATA HUMAN ALPHA HAEMOGLOBIN (FIRST 40 AMINO ACIDS )
1001 DATA V LSPADKTNVKAAWGKVGAHAGEYGAEALERMFLSFPTT
```

```
1010 DATA HUMAN MYOGLOBIN (FIRST 40 AMINO ACIDS )
1011 DATA G LSNGZWE VLNVWGKVEPNIAGHGEEVLIRLFKGHPET
2000 DATA END
```

The main part of the program begins at line 32 when the name of the protein to be investigated is read into the string variable N$. Since this is the first READ statement the string on the first DATA line (1000) is assigned to N$. Note that it is not necessary to enclose the string within double quotes on a DATA line. This string is compared with "END". If the two are equal the program has run out of data and hence the program terminates. If the program has not ended the next string, which is an amino acid sequence, is assigned to the string variable S$. The user is then asked to enter the string whose frequency of occurrence in S$ will be investigated. This is assigned to the string variable A$.

The search for the occurrences of A$ in S$ is completed by the loop on lines 54 to 58. Note that A$ can be of varying lengths including a single letter. This facility could be used, therefore, to determine the frequency and positions of particular amino acids in a range of proteins. The search algorithm is based on a simple analysis of S$. A$ is compared with segments of S$ starting at the first position and then moving, one position at a time, to the final amino acid. In order to do this the loop must be executed as many times as there are amino acids in the protein. This number is found by setting the end value of the loop to be equal to the length (LEN) of S$. The comparison between the S$ segments and A$ are carried by the instruction on line 56. The MID$ function isolates a fragment of the string S$ which starts at position I and is as long as the string A$. If a match is found between this fragment and A$ its position is printed as the current value of the loop counter I.

Lines 60 to 71 are concerned with offering various options to the user. These options make use of the INKEY$ function (which could be replaced by GET Q$). Line 62 asks the user to press E to terminate the program. When a key is pressed it is assigned to the temporary string variable Q$. The contents of Q$ are then compared with E. If a match is found the program ends, if not the program continues and line 66 is executed. This requests the user to press either P or S depending which option is required. Line 70 examines the contents of the string variable Q$. If the contents of Q$ is not P or S it is rejected as an invalid character and control returns to the keyboard scan routine on lines 68 and 69. Line 70 selects which of the two options will be carried out depending upon the contents of Q$.

Note
The blanks in the amino acid sequences of alpha haemoglobin and myoglobin mark positions where it is believed that an amino acid has been deleted. This is based upon comparisons with other proteins from the same family.

7.5 Summary of commands

String variables   A string variable is characterised by the terminating $ symbol. All other rules concerning the structure of variable names apply. The range of characters allowed for a string variable is generally 0 to 255. Most dialects also allow subscripted string variables which must be dimensioned as in numeric arrays. The following are all valid string variable names: T$ W1$ L$(3) Q$ Z9$(I)

ASC (string)   This command returns the ASCII code of the first character in the string argument.

CHR$ (number)   The character, whose ASCII code is the numeric argument, is returned.

VAL (string)   VAL converts a string of numeric characters into the equivalent number.

STR$ (number)   STR$ converts a number into its equivalent character string.

LEN (string)   This command returns the length of the string argument.

LEFT$ (string,X)   LEFT$ returns the X leftmost characters of the string argument.

RIGHT$ (string,X)   This is equivalent to LEFT$ except that the X rightmost characters of the string argument are returned.

MID$ (string,P,N)   MID$ returns the string of N characters from the string argument which begin at position P.

GET or INKEY$ These are equivalent commands which are not usually implemented together. They scan the keyboard and assign the character of any key that is pressed to a variable. The RETURN key is not used with this form of input. Examples:

GET Q$
Q$ = INKEY$

PRINT USING "#s";   If available this command allows a wide range of data formats to be established. The only use of this command covered here concerned the formatting of numbers. The number of characters before and after the decimal point are set by using # characters. A . is used to mark the position of the decimal point.

7.6 Problems

1. Modify the data entry routine in section 7.4.1 so that it will accept decimal numbers. The ASCII code for . is 46. You should also include

an additional error check which will prevent more than one . from being entered into one string.

2. Write a section of code which could be substituted into the amino acid sequencing program so that two proteins can be compared. The program should count the number of identical matches, i.e. the same amino acid in the same position.

3. Further modify the amino acid sequencing program so that the protein names and amino acid sequences are read into two string arrays. It should then be possible to provide the user with a facility for selecting, by name, which protein is to be investigated.

4. The simple statistics program, Example 6.1, can be improved by including units in the output. Add the additional code which is required to input and output the units as appropriate.

5. Finally, write another section of code which could be used with the amino acid program to count and tabulate the frequency of each amino acid in a protein. This can be achieved by a relatively simple loop which makes use of the fact that the ASCII codes for the letters A to Z have sequential values between 65 and 90. Note that the letters J, O, U and X do not form part of the code. You will need to exclude these from the loop.

Chapter 8 Disks, Files and Operating Systems

8.1 Introduction

These three apparently diverse topics are in fact closely related. Large computer files can only be implemented successfully on computers which have access to disk storage systems and disks will not operate without a disk operating system.

8.2 Disks

8.2.1 Disk organisation

A disk is a magnetic device which is used to provide backing store. Physically it resembles a gramophone record, in that it is a flat disk which is covered in circular tracks. The tracks do not spiral as on a gramophone record but instead form a series of concentric circles. The disk is composed of either light alloy or plastic which is overlain by a magnetic material. The information is recorded onto (written) and read from the disk by one or more read/write recording heads which are similar to those found in tape recorders. All disks, but not the disk emulators, operate in similar ways. A disk emulator is a storage system which does not use a disk but mimics the disk interaction with the user. The disk recording surface is broken down into three components, the first of which are the concentric rings or tracks. The number of tracks on a disk varies with disk type. Each track may be thought of as a piece of recording tape which has been formed into a circle. The tracks are divided into sectors, the second component. The sectors are marked by a number of 'spokes' which radiate out from the disk centre. Again the number of sectors differs between disk types and model of computer. Each sector of a track is known as a block and forms the last component. A block normally holds a constant number of bytes of information, usually a multiple of 128. Since the outer tracks have a greater circumference than the inner tracks the recording density changes with position on the disk. The spaces between the sectors are used to mark the positions of the sectors and are not used, therefore, for the storage of data and programs.

Since the information on a disk is organised into discrete blocks, which can be identified by their track and sector position, all of it is equally accessible. The disk operating system maintains a directory which contains the position of each separate file of information. If access to one of these files is required the read/write head is moved over the appropriate track and it then waits for the correct sector to come under the head. The information in that block can now be accessed by the computer. Since all blocks are equally accessible the disk system is said to be a random access storage device. Tape systems are sequential access storage devices since information can only be accessed in the sequence in which it was recorded.

Before a disk can be used the disk surface must be organised into a format which is suitable for the computer, a process known as formatting. Formatting consists of marking the tracks and sector positions magnetically. Older disk systems use hard sectoring to format the disk surface. If a disk has hard sectors their positions are permanently marked by a number of holes punched close to the disk centre, thus the number of sectors is fixed. Most modern disk systems use soft sectored disks which have one hole punched near to the centre. This is a reference point which marks the start of the first sector. Other sectors can then be marked magnetically. This allows the same disks to be used by computers which utilise different numbers of sectors per disk. It also creates the situation in which disks prepared on one type of computer cannot be read by a different type of computer.

The organisation of a soft sectored disk surface, into separate components, is shown in Fig. 8.1.

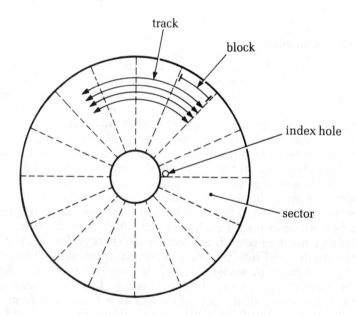

Fig. 8.1 Diagram of a 16 soft sectored floppy mini disk (with protective cover removed)

Three main types of disk can be recognised: 'floppy' disks, hard disks and disk emulators.

8.2.2 'Floppy' disks

'Floppy' disks get their name because of the flexible plastic material from which they are constructed. It is not a good idea to investigate this property since bending a 'floppy' disk may damage the disk surface and hence destroy any information stored there. Because the disk surface can be damaged easily it is normally protected by a non-removable jacket which has openings to allow access by the recording head. When a disk is placed in a disk drive the centre of the disk is held between a pair of rings which are connected to the drive motor. When the disk is spinning, at speeds of up to 300 revolutions per minute, it will rub against the jacket interior. Even though the jacket interior is coated with low friction material there will still be some wear on the disk surface. It is for this reason that 'floppy' disks should not be thought of as permanent backing store devices. The disk can also be damaged by heat and magnetism. It is very good practice to prepare 'back-up' copies of any material stored on disk. These back-ups should then be stored away from the originals.

'Floppy' disks are now available in a variety of formats. The variables to be considered are the disk diameter, the recording density and the number of disk surfaces used for recording.

The original 'floppy' disks had a diameter of eight inches. Typically they have 77 tracks with 26 sectors per track. Each block has a storage capacity of 128 bytes. If only one side of the disk is used this gives a storage capacity of $77 \times 26 \times 128$ bytes (256,256 bytes). This would be a single-sided, single density disk. Improvements in disks and disk drives have allowed information to be stored at greater recording densities. This means that 256 bytes can be stored in the same physical space as the 128 bytes on a single density disk. A disk which uses this doubled recording density is called a double density disk and could hold $77 \times 26 \times 256$ bytes (512,512 bytes). If both sides of the disk are used for recording, by having two recording heads, then the capacity will be doubled to over one million bytes. A disk which has information recorded on both sides is a double sided disk.

The next generation of 'floppy' disks were the 'minifloppies' which have a diameter of five and a quarter inches. One important difference between these and the eight inch disks is that the latter are spinning all of the time that they are in the disk drive whereas the 'minifloppy' disks only spin when data is being read from or written onto the disk. The amount of information which can be stored on a 'minifloppy' depends upon the recording density and the number of sectors and tracks (typically 40 or 80 tracks). It is possible to have a range of storage capacities, on the same disk, of between 88,000 and 1,200,000 bytes depending upon the type of computer and operating system used. A new generation of 'microfloppies' has been introduced recently. They are available in a variety of sizes, three, three and a quarter and three and a half inches. The main difference between these and the other types of 'floppy' disk is size.

8.2.3 Hard disks

As the name suggests these disks are made from a rigid material which is usually a light alloy. As with the 'floppies' this base material is covered by a layer of magnetic material which is arranged into a number of tracks. Hard disks can store much more information than 'floppy' disks. One surface of a hard disk can store between one and thirty Mbytes (1 Mbyte = one million bytes) depending upon the number of sectors and tracks which are available. A hard disk will have between 300 and 1000 tracks on one surface. This vast increase in data packing is a result of the much more sophisticated design of the read/write head. In a hard disk system the recording head is positioned about 50 millionths of a cm above the disk surface. Normal mechanical devices cannot achieve this precision, therefore a different principle is used. When the disk spins, at over 3000 revolutions per minute, it creates turbulence. The recording head is fitted with a small wing which produces lift when the disk is spinning and creating air movements. If the recording head touches the disk surface it will destroy any information at that point and so produce a disk 'crash'. When the disk is stopping the air movements will decrease and the lift produced by the wing will be lost. Surface damage is prevented in these circumstances by positioning the head over a special empty track which never carries information.

Data transfer rates to and from hard disks are much faster than for 'floppy' disk systems. This is because the hard disk is spinning over ten times faster and hence a block of data will arrive at the recording head more quickly. In addition the recording head moves into position much faster. Both of these improvements produce a five-fold increase in data transfer rates over 'floppy' disk systems.

The hard disks may be fixed or removable. Because a hard disk would be ruined if it was contaminated by dust particles (which have a larger diameter than the gap between the disk surface and read/write head) those systems intended for home or business use are usually sealed for life. The disks cannot be removed. This type of hard disk system is known as a Winchester after the IBM code name for the project which developed them. If a removable hard disk system is used then steps must be taken to ensure that the atmosphere around the disk is free of dust. This increases the cost of these systems.

The storage capacity of a hard disk drive can be increased by mounting several disks on a central spindle and providing a read/write head for each disk surface. These are known as multiplatter drives. The storage capacity of a hard disk system will normally be around 5 to 10 Mbytes for a Winchester drive used with a microcomputer, with a maximum of about 80 Mbytes. The hard disk systems used on a mainframe computer may have storage capacities of over 2.5 Gigabytes (1 Gigabyte = 1,000,000,000 bytes).

8.2.4 Disk emulators

A disk emulator is a storage device which uses disk commands and carries out normal disk operations but without using a disk. The most useful of these is a RAM disk. This is a block of random access memory, typically 256K, which is used as a disk drive! A directory is established which can be used to locate 256 byte blocks of data which are stored in the RAM disk. Since no mechanical operations are required to access data in the RAM disk the data transfer rates are much faster than normal disks. The information in a RAM disk will be lost when the power is switched off. Therefore, the information is transferred to a real disk when the session has been completed. The other main type of disk emulator uses a closed loop tape system. This kind of device, which is known as a floppy tape system or microdrive, is in a fact a sequential access storage device. However, since the tape is joined to form a loop it can continuously cycle past the read/write head. Even though the data is transferred at much higher rates than with domestic tape recorders, floppy tape systems are much slower than true disk systems.

8.3 Files

8.3.1 Introduction

Files are a major component of most useful computer operations. Various types of file can be recognised including textfiles, program files and datafiles. As its name suggests a textfile contains text. This could be produced by a word processor or it could be the source code (an uncompiled program) for a Pascal or FORTRAN program. A program file would contain a program which could be executed. This section will, however, be restricted to the final type, the datafile. A datafile (henceforth referred to as a file) is an organised collection of information which relates to a common base. Biological examples would include a file of protein amino acid sequences, a file of life history information about a group of experimental animals, a catalogue of drugs and information about their structure, operation and interactions. These files would all be permanent, although they could be updated as required. Computers also use temporary files. These may be used to hold intermediate data during the execution of a program. Most non-trivial programming will require an understanding of file processing and organisation. Meaningful file operations can only be carried out if disk storage is available even though tape cassette file handling commands are provided in many BASIC dialects.

8.3.2 File structure

The information contained within a file must be carefully organised if it is to be of any value. This can be achieved if the file is arranged as a collection of sub-units. This type of organisation will be described with an example. Consider a laboratory which has a large number of experimental animals. The laboratory will require access to information about each animal, for example, date of birth, strain number, sex, etc. The complete collection of information about all of the animals would constitute a file. The file can be initially subdivided on the basis of information about individual animals. Each animal would represent one record within the file. A file consists, therefore, of a group of records which contain information about some common element, in this case experimental animals. Each record can be further subdivided into fields of information. In this example the fields would consist of information about the age, sex, strain, etc. of each animal. Each record should be broken down into the same set of fields arranged in the same sequence. Table 8.1 illustrates this type of file organisation.

Table 8.1 File organisation

Filename: Exptanimal

Record no.	Animal no.	Sex	Date of birth	Strain	Cage no.	Expt reference no.
001	13654	M	10/09/83	12B	103	345C
002	09932	M	11/08/83	12B	028	000
003	15211	F	03/11/83	09F	121	345C
004	11190	M	21/01/84	14D	238	112E
.	.	.	.	.	.	.
676	07742	F	07/08/84	09F	566	113C

The file in Table 8.1 has been given the name *Exptanimal*. Each computer system will have rules which control the type of filename used. In many systems the filename may be restricted to eight or fewer letters and often filename extensions must be given. A filename extension is used to describe the file type, again the type of extension used will be determined by the computer system. Examples of file extensions would include .BAS for a BASIC program file, .FTN for a FORTRAN program and .$$$ for a temporary file.

The Exptanimal file is composed of 676 records (not all shown!). This laboratory currently has 676 animals in stock. Each record is broken down into seven fields: the record number, the animal number, sex, date of birth, strain, cage number and experimental reference number. The fields can be characterised by their width (number of characters) and data type. The first field, the record number, has a field width of three and the data is numeric. The second field is also numeric but is wider at five characters. The third field, the sex, is an alphanumeric field one character wide. The fourth, fifth and seventh fields are also alphanumeric since they contain characters which are not numbers.

Once the file has been created it can be used for various purposes. In particular it provides a source of information about individual animals. If additional fields, such as weight and diet, were included the file could be used to extract information about the growth rate and performance of the animals. The use of files as sources of information is discussed in Chapter 12.

The Exptanimal file structure shown in Table 8.1 is not the only one available. In fact this file structure would make it difficult to search the file for specific information. It would, for example, be quite time consuming to determine the cage number of animal number 07742 (the last record). File searching can be improved by using another type of file structure in which the information is sequenced or indexed. None of the record fields in the Exptanimal file is ordered, the records having been added to the file with no apparent organisation. There are various ways in which it could be reorganised. One possibility is to arrange the records in an ascending animal number sequence. Record one would be occupied by the lowest numbered animal, record two by the next highest numbered animal, etc. Alternatively, the file could have been arranged in ascending cage number order. Other types of ordered structure would be equally valid.

The field which is used to provide the file structure is known as the key, and file ordering or indexing relates to this key. This aspect of computing can be very complex and only a simple introduction can be provided here. There are three main methods or algorithms which can be used to aid file searching. These are sorting, pointing and hashing.

If a sorting algorithm is used the file must be resequenced, using the key field, every time that the file is changed or updated, for example by including new animals or changing some of the fields within a record. This operation can become very unwieldy and time consuming on large files even when efficient sorting algorithms are employed.

The pointing algorithm allows the records to be entered into the file in any sequence. However, each key field, more than one may be used, has two pointers associated with it. These pointers contain the record numbers or addresses of the previous and next record. Consider the list of five animal numbers below, taken from Table 8.1.

Record no.	Animal no.
1	13654
2	09932
3	15211
4	11190
5	07742

These numbers are not in a numerical order. The lowest numbered animal is the last record, the next animal is the second record, etc. These relationships can be included in the file as pointers as shown below.

		Pointer to record	
Record no.	Animal no.	Above	Below
0		5	0
1	13654	3	4
2	09932	4	5
3	15211	0	1
4	11190	1	2
5	07742	2	0

An additional record, record number 0, has been added. This has no information in the animal number field but in the first pointer field it contains the record number of the lowest numbered animal. This is animal number 07742, which is held in record 5. Since there are no records with a lower number, the below pointer in record 0 points to itself. The pointers attached to record 5 indicate the positions of the next highest and lowest records. The next highest record is record 2, 09932. The pointers for this record point to records 4 and 5 as the animals above and below number 09932. Each time that new records are added the pointers must be changed to take account of the new sequence of key fields. One advantage of this technique is that a file can be indexed (sequenced) on more than one key by establishing pointers for several fields.

95

The final technique to be described here is the hashing algorithm. At first this appears to be a clumsy method, but when it is implemented on long files it can become one of the most efficient file indexing techniques. This technique also indexes files with reference to key fields but neither sorting nor pointers are employed.

Initially a table is created which has more entries than the potential number of records. The key field entry is then subjected to a mathematical manipulation the result of which determines its table position. If the field is alphanumeric the ASCII codes for the symbols can be used to derive a number for the manipulation. One of the commonest operations is to divide the key field number by the number of entries in the table. The table position occupied by a record will be that of the remainder from the division. Consider the Exptanimal file. The file is to be indexed by the animal number. Initially a table of 1000 entries is established. Therefore the table entry for record 1 will be position number 654 (13654/1000 = 13 with a remainder of 654). Record 2 occupies position 932 (9932/1000 = 9 with a remainder of 932). Other records would be allocated table positions in a similar manner. When information about a particular animal is required the key field number is subjected to the same mathematical manipulation. This would yield its table position. For example, if information about animal number 13654 is required and its file record number is not known it can be quickly determined by using the hashing technique. 13654/1000 leaves a remainder of 654. The contents of position 654 would be the animal number 13654 and its record number which is 1. In this way individual records can be very rapidly accessed.

You may have already realised that there are problems with this method. Any animal number ending in 654 would yield the same table position, i.e. collisions will occur. Various algorithms are available to deal with this inevitable problem, the simplest of which involves placing a record in the next available table entry. This means that a linear search of the table, away from a record's ideal point in the table, will be required when a table entry does not contain the specified field key. Even so the search will usually be very brief. More complex algorithms set recommended table lengths. It has been demonstrated that table collisions will be reduced if the table length is set according to the following rules. The table length should be a prime number which is equal to (4 × an integer constant) + 3. It should, of course, be longer than the potential number of entries. In the above example a better table length would have been 1019. This is a prime number which is equal to 4 × 256 + 3. Animal number 13654 would occupy table position 507 (13654/1019 = 13 with a remainder of 507). Animal number 14654 would be in position 488 (13654/1019 = 14 with a remainder of 488).

8.3.3 File handling commands

Most high level languages contain a set of commands which can be used to manipulate datafiles. Even though there are the inevitable differences between languages and dialects there is a degree of similarity between the commands used. Before a file can be used it must be opened, hence the command OPEN. This is usually combined with additional information such as the filename and number and whether it is to be used for input or output. When file transactions have been completed the file should be closed with the command CLOSE. If a file is not closed correctly it will usually lead to the loss of information since the final pieces of information may not have been transferred from a memory buffer to the file. Information is exchanged between a file and the computer by a pair of commands such as GET and PUT or INPUT and PRINT. The use of the PRINT command may seem strange. It is used because the disk can be treated as an output device onto which information can be printed.

BASIC allows two types of datafile to be created. These are sequential and random access files. Records in a sequential file can only be accessed sequentially. This means that all previous records must have been written or read before a record can be accessed. If the file is a random access file any record can be accessed and worked upon independently of all others within the file. It is much simpler to write programs for the less useful sequential access files.

The main difference between these two filetypes is in the record length. Whereas sequential files can have records of different lengths, in a random access file the record length is fixed by the user or the operating system. Since all records in a random access file contain the same number of characters, it is a relatively simple operation for the disk operating system to identify the position of a record on the disk surface if the position of the first record is known. A single record can, therefore, be identified and transferred into main memory where it can be operated on as required.

Note
Both of these file types can be organised by using the sort, pointer and hashing algorithms described in the previous section (8.3.2).

8.4 Operating systems

8.4.1 Introduction

If operating systems did not exist very few people would be willing or able to use computers. An operating system acts as an interface, which is often transparent to the user, between programs and the computer hardware. An operating system is also a program which is written in machine language and may be present in ROM. The simplest operating systems are those found in inexpensive home computers and are often known as monitors because they monitor and control the hardware. They control and manage inputs from the keyboard and audio cassette recorders. The monitor programs also manage the output of material to the output device. BASIC system and program commands may make calls on subroutines present in the operating system which in turn control the requested operation. If operating systems did not exist it would be necessary to write subroutines which would control operations such as printing the appropriate character at the correct position on the VDU screen.

The more complex operating systems are related to disk usage. These are known as Disk Operating Systems (DOS). The exchange of information between a computer and a disk requires a large number of complex operations to be carried out in the correct sequence and at the correct times. These complex disk operations may be controlled by issuing a single command. Consider the problem of determining which files are present on a disk. Before this information can be obtained the following operations must be completed.

1. A communication channel must be opened between the computer and its disk drives.

2. The drive motor must be started.

3. The read/write head must be positioned over the track which carries the directory information.

4. This information must be read from the disk in the correct sequence before it is transferred to the computer.

5. The read/write head should be returned to its inactive position and the drive motor stopped.

6. Once in the computer the directory information must be decoded and printed onto the output device.

7. If an error should occur (for example, there was no disk in the drive or an unformatted disk has been used) the nature of the error should be communicated to the user.

All of these operations are under program control via the operating system. The operating system will normally be provided with a range of commands to control these operations. In Apple DOS the correct command would be CATALOG, while other systems would use DIR, or a similar command, for the same operation.

Until quite recently microcomputer disk operating systems were quite simple with, in general, each computer having its own system and commands. One exception to this was the CP/M operating system developed by Digital Research Inc. CP/M was written to be used with a family of CPUs rather than individual computers. This had an important effect on the software market since programmers could write programs, which utilised the CP/M operating system, to run on a variety of computers. This vastly increased the potential market for a program and hence the number of programs written to run under CP/M control. Since the introduction of 16- and 32-bit microcomputers there has been a vast improvement in the power and flexibility of disk operating systems. Some of these are scaled down versions of operating systems that were originally written for mini- and mainframe computers.

8.4.2 Operating system features

The main features of any comprehensive operating system are concerned with easing input/output operations and file manipulations. A good operating system will contain programs, usually called device drivers, which are used as a software interface between the computer and its input/output peripherals. They function by converting data from its internal format to that required by the peripheral. An operating system normally recognises a small number of logical devices, for example, a list device. However, each logical device can have a variety of physical devices assigned to it. Those physical devices which fulfil the list device logical functions include CRTs and printers. As long as the operating system has the appropriate device driver program it should be possible to redirect the list device output to the selected physical device.

The second major feature of an operating system is its file handling commands. These will include utilities such as directory listing, deleting, copying, merging and renaming of files. All of these operations should be performed in response to single word commands (which inevitably differ between operating systems). When a command is issued the operating system will look through a directory of executable files (programs which can be run) and if a match is found between the command and a file the program will be started. All of the file operations are under the control of programs which form part of the operating system. There is, therefore, a relationship between the power of an operating system and the amount of disk or memory space that is required to hold it.

A vital part of the operating system is the editor, which is used to create and amend textfiles. It is not usually possible to use an editor to create binary files. An example of a binary file would be an executable program written in machine language. If the contents of a binary file are displayed the screen will usually fill with various flashing random characters, because such a file was never intended to be viewed in this way. The difference between a textfile and a binary file is a reflection of the fact that textfiles are normally created by humans, whereas binary files are normally created by the computer.

The most recent advances in microcomputer operating systems have been associated with the increasing capacity of disk storage systems and CPU performance. These advances are highlighted by the multiuser, multi-tasking operating systems which are now available. A multiuser operating system allows more than one user to use the same CPU. This does not mean that several people are attempting to use the same keyboard and screen. Remember that a computer system is composed of several discrete units which are centred on the CPU. The combination of screen and keyboard can be thought of as a dumb terminal since they have no computing power of their own. Several of these dumb terminals can be connected to one 'black box' which contains the CPU and internal and external memory. Obviously this arrangement will require careful management, by the operating system, if chaos is to be avoided.

There are several approaches to the problem of managing a multiuser system. The simplest, but least satisfactory, is first come, first processed. This would mean that access to the CPU would be denied if another user was already connected. Another solution is to divide the CPU time equally between the users. If this system is adopted each user is allocated a fixed time period of CPU time. When this period expires the CPU stores its intermediate results in a temporary file which will be recovered when the next time period comes round. Although this sounds a democratic solution it is not very efficient. The management of a multiuser system can be improved by establishing priorities and maximising the efficiency of information input and output. This last point is particularly important since the CPU is often idle while a program is carrying out an input or output. While one program is thus employed another program, which requires computation only, could use the CPU.

Another problem with multiuser systems is that several users may be connected to the same datafile. This can create problems if a user tries to access or delete a file which is already being worked on by another user. If access is granted, the file will rapidly become a mess as users make different changes to the same information. This situation is avoided by a process of record-locking; data which is currently being used is denied to all other users.

An analogous feature of some operating systems is that they are single-user, multitasking systems. This indicates that one user can have several programs running at the same time, or concurrently. For example, four programs may be run concurrently by allocating CPU time to the programs as required. This process can be established because the operating system recognises four terminals, one real and three imaginary or virtual terminals. CPU time is allocated by using similar principles to those outlined for the multiuser systems. The user can move between the terminals by pressing the appropriate control keys. The screen will display the output of the program running on the currently attached terminal. One impressive feature of some of these systems is that parts of the screen can be allocated to the four terminals so that all four programs can be monitored simultaneously. This technique of splitting the screen into independent areas is called windowing.

The most 'talked about' new generation of microcomputer operating systems is the Bell Laboratories UNIX system. This was written for powerful minicomputers such as the DEC VAX 11 series. It is an extremely powerful and flexible system which is now being implemented on many of the Motorola 68000 based microcomputers. UNIX is a true multitasking, multi-user operating system which provides many powerful facilities including tree directories, print spooling, record locking and passwording. An extra feature of UNIX is that it has very powerful input and output routines which use software tools called pipes and filters. A pipe can be used to direct the output of one program into the input of a different program, without the use of intervening temporary files. It is possible to set up a system so that data from one program passes through a number of other programs, each one altering the data in some way. These programs are the filters. This type of operation is not possible with most microcomputer operating systems. UNIX is a large operating system and when implemented on a microcomputer requires a 68000 microprocessor and a Winchester of at least 5 Mbytes capacity.

An important feature of the UNIX, and some of the other modern microcomputer operating systems, is that it uses a tree directory system. The large storage capacity of a hard disk, combined with a multiuser operating system, can mean that the disk directory contains hundreds of files. It is important that this directory is structured in some way if both the user and the operating system are to make efficient use of the files. Imagine trying to find the name of one of your files when it is surrounded by several hundred others! The operating system will normally use a hashing technique (see section 8.3.2) to locate the files. The user is allowed to construct tree directories. In a tree directory some of the entries are allowed to be

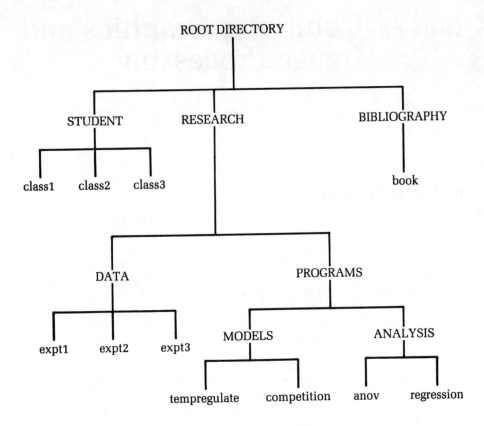

Fig. 8.2 Example of a tree directory. Directories are shown in uppercase.

subdirectories which contain information about files on the next level. This type of file directory is very similar to the familiar phylogenetic classification schemes used by biologists. A tree directory is illustrated in Fig. 8.2.

Individual files are referenced by their pathnames, i.e. the route which must be followed from the root directory. Thus, in Fig. 8.2, the pathname for the file 'competition' would be ROOT/RESEARCH/PROGRAMS/MODELS/ competition. Files can have the same names as long as they reside in different subdirectories.

It is almost certain that the operating systems which will become available for the next generation of microcomputers will have more in common with those found currently on mini- and mainframe computers rather than those available for the 8-bit microcomputers.

Chapter 9 Computer Graphics and Image Processing

9.1 Introduction

Computers are becoming increasingly important tools for the production of visual representations of a wide variety of subjects. Two related areas can be recognised, both of which have many existing and potential biological applications. The first, computer graphics, is concerned with the construction of images from mathematical information about their shape and structure. Computer graphic techniques can be used to construct various pictorial representations of data, for example, graphs, histograms and pie charts. The same techniques can also be used to produce two and three dimensional images of a variety of biological structures ranging from molecules to skeletons. The second major area is image processing. This technique begins with an image which is analysed (a mathematical description of the image is produced) and then reconstructed before being displayed. The reconstruction techniques are used to remove unwanted information ('noise'), to enhance the contrast between adjacent structures or to search for patterns in the images being analysed. These image processing techniques can, therefore, be very useful techniques for refining the images of microscopic structures such as chromosomes or analysing satellite photographs of the earth's surface. Section 9.4 will review some of the existing biological applications of both of these techniques.

There is only one example program in this chapter. This is because graphics commands are generally very machine specific. Even commands which are superficially very similar can have quite different effects on different computers.

9.2 Hardware review

Both computer graphics and image processing are very dependent upon specialised hardware for many of their applications. This section will introduce, in a non-technical manner, some of the relevant hardware.

There are three main hardware components in a computer system which is used for graphics. They are display devices, input devices and specialised processors. Only the first two will be discussed here.

9.2.1 Display devices

There are two main categories of display device. The first category consists of devices which utilise Cathode Ray Tubes (CRT), while the second is composed of a wide range of printers.

The CRT devices can be further subdivided on the basis of how information is written to the CRT screen. Before these systems can be described it is necessary to understand a little of how images are generated on CRT screens. The inside surface of a CRT screen is coated with phosphor. This material will emit light when it is bombarded with electrons. The electrons are generated by an electron gun, the cathode. The intensity and direction of travel of these electrons is controlled by various components within the CRT. The number of electrons striking the phosphor-coated screen can be varied by altering the voltage to a control grid which is placed between the cathode and the screen. This is important because the intensity of light emitted by the phosphor is determined by the number of electrons which strike it. The light emitted by the phosphor normally decays when the supply of electrons is interrupted. The rate of decay will depend upon the screen design, but will generally be in a range between 10 microseconds and 10 seconds. If the light decays rapidly the image produced by the glowing phosphor must be quickly redrawn otherwise the image will appear to flicker. The time taken to redraw the display depends upon which of the two main designs is used. If the image is retained for a reasonably long period of time this can create problems when moving images are required. The old image will persist even after the next frame of movement has been drawn.

Most biologists should be familiar with the concept of resolution as applied to microscopes. Similar criteria can be used to evaluate a CRT display. If a display is high resolution it should be possible to distinguish between two spots which are very close together in the display. The degree of resolution is also related to the mode of action of the CRT.

The highest resolution displays are usually found on systems which generate images in a different way from that used in television sets. The earliest computer graphics displays made use of these systems called direct-view storage CRTs (DVST). These are still found in systems in which a very high degree of resolution is required. The DVST is different from a normal television set in that once a picture has been written onto the screen it will remain there until erased. This is achieved by using a different design principle. Displays which are generated on DVSTs are restricted to lines and the display can only be one of two possibilities: on or off. Dynamic (moving) displays, shading and the filling in of colour blocks are not possible. The display is stored by the computer as a set of coordinates. A line is therefore defined by its start and end coordinates. The controlling hardware and software will draw a straight line between two coordinates to produce the line. This kind of plotting is known as vector plotting. The display can be built up in any sequence. All that is required is the start and end coordinates of all lines that must be drawn. A related technique is the stroke/refresh tube. The screen image is not permanent and will decay with time. The image is redrawn (refreshed) by using vector plotting techniques. The screen coordinates which are needed for the refresh are stored in a short term memory called the refresh buffer. Stroke/refresh tubes may suffer from a flickering image if the display is very complex. This is because the early parts of the image may start to fade before the last parts are drawn.

Most microcomputers use raster scan display techniques. A domestic television set is a raster scan display. In such a display a line is not defined by its start and end coordinates but the entire line is stored in memory as a set of points. The display produced by a raster scan technique is made up of a series of points or pixels (picture elements, also known as pels). These are written to the screen by a fast moving electron beam which scans from left to right, starting at the top, and quickly moves down the screen in a sequence of left to right passes. When the beam reaches the bottom right it quickly moves back to the top left and starts to write the next frame of information to the screen (between 30 and 50 times per second). A pixel may be illuminated by varying the beam intensity. Unlike DVSTs and stroke/refresh tubes, information about the status of each pixel must be stored in memory for a raster display. In the simplest situations a pixel may be either on or off. It is therefore possible to store information about a pixel in one bit. This type of display is similar to the stroke/refresh system in that the light emitted by the phosphor decays quite rapidly (this is necessary for moving images). Flickering of the image can only be avoided if the image is quickly redrawn. The information which is needed to reconstruct the display is stored in a refresh buffer. If a line is to be redrawn in a new position all of the pixels which make up the old line must be set to off while all those pixels on the new line must be set to on. This is a much less efficient use of memory than the DVST and stroke/refresh systems in which only two coordinates need to be changed. The relationships between these three types of display are summarised in Table 9.1.

Table 9.1 Relationship between different types of display

Phosphor image decays	Method of writing to the display	
	Vector plotting	Raster scan
yes	stroke/refresh	domestic television type
no	DVST	not applicable

It is not possible to use domestic television sets for the display of high resolution graphics. This is because the maximum number of pixels that can be displayed is approximately 300 by 585 (157,500 pixels). If colour is used the maximum resolution is even lower, about 160 by 585 pixels. High resolution graphics is only possible with video display monitors which have a wider video band width. The band width determines how many individual points can be turned on or off during one passage across the screen. Domestic televisions have a band width of about 5.5 MHz while monitors frequently have band widths of 20 MHz. They are, therefore, capable of controlling approximately four times as many pixels as domestic televisions. Monitors also have four times as many phosphor dots to take advantage of this facility. The resolution of a 20 MHz monitor is about 800 by 800 pixels (640,000 pixels).

High resolution graphics places a very large burden on a microcomputer using a raster display. In a monochrome display each pixel requires at least one bit of memory. A medium resolution display of 256 by 256 pixels requires 256×256 bits of memory. This is 65,536 bits or approximately 8K of memory. If colour graphics are being used the memory overheads become even more severe since several bits will be needed to carry the additional information about each pixel colour. Note that as the choice of colours increases so does the number of required bits. The highest resolution raster displays of 1024 by 1024 pixels require over a million bits of memory! Because vector plotting systems do not need to store information about the condition of every screen pixel they can be used with much higher resolution displays without imposing the same heavy load on the memory. High resolution DVST displays may have a resolution of 4096 by 4096 pixels (equivalent to 2 Mbytes of raster display memory). A high resolution raster scan display creates problems other than memory requirements. Any changes to the display must be written into the memory before the next frame is displayed, about 1/10 of a second. This is beyond the capabilities of most present microcomputers. Despite these problems raster displays do have advantages over vector displays in that it is possible to have moving images and blocks of colour.

Hard copies of computer graphic displays can be produced on a variety of printers. The normal type of dot matrix computer line printer used with microcomputers can usually produce only quite crude copies. The images will be of poor quality because the printhead will have a restricted movement and the resolution will be controlled by the number of pins which make up the printhead. This type of printer can be considered to be the equivalent of a raster display CRT. There are also printers equivalent to the vector plotting displays: these are the printer/plotters. The image is produced by moving a pen in a combination of horizontal and vertical directions; thus a picture can be generated in any sequence of lines. The resolution of these plotters will be determined by the magnitude of the smallest possible horizontal or vertical pen movement. This type of printer is generally slower and more expensive than dot matrix line printers.

9.2.2 Input devices

One of the most useful advances in computer graphics technology has been the development of interactive displays. This allows the user to control the display, within certain limits, without investigating the controlling software. There are a number of specialist devices which ease the interaction between the user and the program. The most primitive of these devices is the keyboard which may have certain keys programmed as function keys. If programmable function keys are available they can be used to control functions such as scaling and rotation of the image. The more specialist input devices include equipment such as data tablets, mice, joysticks and light pens.

A data tablet is an electronic device which can be used to input x,y coordinate information in a very simple manner. The data tablet is a board over which a stylus or hand cursor can be moved. Various techniques are used to detect the position of the stylus or cursor on the tablet. If a drawing is placed on the tablet then its outline can be traced and converted to a screen image. The information is obviously transferred to the computer as a set of coordinates which are then used to generate the image.

The mouse is becoming a common component of the more expensive business orientated computer systems. A mouse is a hand-held locator which can detect horizontal movement in two orthogonal (at right angles) directions by virtue of two sets of wheels set at right angles to each other. The movement of the mouse on a flat surface is mirrored by the movement of a cursor across the screen. The mouse is usually equipped with additional input devices in the form of push buttons. These can be used to send simple signals to the computer.

A joystick is a lever which can be moved through an angle of 360 degrees. The direction of movement is detected by a pair of potentiometers which record the amount of movement through two orthogonal axes. This movement is transferred to equivalent movements of a screen cursor. Joysticks are difficult to use with precision because the length of the lever produces an amplification of even small hand movements. It is possible to have joysticks in which the lever can be twisted. This provides for movement in a third dimension which is useful with three dimensional images.

The final input device covered here is the light pen. A light pen uses a photoelectric cell to detect light. The photoelectric cell is usually insensitive to the light emitted by glowing phosphor. It is, however, able to detect the higher light intensities associated with the electron beam which produces the phosphor image. If the light pen sends a signal to the CPU when it detects the electron beam the position of the pen on the display can be calculated. It is, therefore, possible to use a light pen for a variety of purposes including selecting options from a menu and moving images around the screen.

9.3 Programming principles

Graphics displays are usually produced by using special graphics commands or software products. It is not usually necessary to write graphics programs from basic principles. Metaphorically this would be equivalent to reinventing the wheel. However, it can be useful to have an insight into the mathematics which underly some of the graphics commands. A very readable and thorough account of the mathematics is provided by Foley and Van Dam (1982). See Appendix E.

Any image can be broken down into a number of points joined by straight lines. This is true even for circles and curves which can be drawn as a series of short straight lines. The quality of the curve will depend upon the number and length of the lines used. If the number of lines is low, a circle will resemble a polygon. A point in space can be defined by a pair of x,y Cartesian coordinates (as on a normal graph) or by its distance and angle from the origin of a baseline. This second system uses Polar coordinates and defines a vector (a vector being a quantity which has both magnitude and direction, distance and angle from the origin of the baseline). These two methods of defining points are illustrated in Fig. 9.1. Most microcomputers use Cartesian coordinates to define points on the screen. This is because the display is usually memory mapped. A memory mapped display is one in which each individual or block of pixels is controlled by a particular memory location. Each of the screen blocks can be identified by a pair of x,y coordinates. Most microcomputers number the top left block with the coordinates 0,0. Unfortunately this creates a few problems with the y coordinates when plotting a graph on the screen. An example of this technique is given in section 9.4.5. (The BBC microcomputer is somewhat easier to use since its graphics screens have the bottom left block labelled 0,0.) The number of separate blocks which make up the display determine the degree of resolution. There is normally a compromise between the display quality and memory constraints. As the resolution increases the amount of free memory decreases.

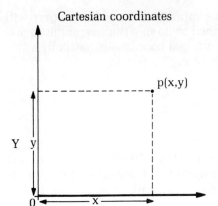  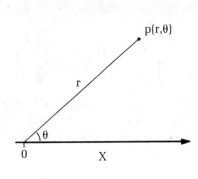

Cartesian coordinates

Polar coordinates

Fig. 9.1 A comparison between the Cartesian and Polar coordinates of a point p

A simple square or rectangle can be defined by the four coordinates of its corners. The simplest way of representing this block of coordinates is as two columns of the x and y coordinates. In this form they can be treated as a matrix. This has several advantages which relate to the mathematics of image handling. The best way to understand the basics of these techniques is to draw a simple structure on a piece of graph paper and go through the mechanics of moving it, changing its size and rotating it. If you do this you will see that there are systematic and predictable changes which apply to all of the coordinates. The coordinate transformations could be calculated individually but a more efficient method is to use matrix algebra techniques. There are standard matrix operations which will carry out even quite complex operations.

Consider the following simple example. A square has four corner coordinates at the following x,y coordinates: 1,1; 1,3; 3,3; 3,1. Try drawing this on a sheet of graph paper. If we wish to move the square to the right and vertically so that its bottom left corner is at 4,2 all of the x coordinates must have 3 added to them while 1 is added to all of the y coordinates. This can be written in matrix notation as:

$$\begin{bmatrix} 1 & 1 \\ 1 & 3 \\ 3 & 3 \\ 3 & 1 \end{bmatrix} + \begin{bmatrix} 3 & 1 \\ 3 & 1 \\ 3 & 1 \\ 3 & 1 \end{bmatrix} = \begin{bmatrix} 4 & 2 \\ 4 & 4 \\ 6 & 4 \\ 6 & 2 \end{bmatrix}$$

The matrix of the initial coordinates has been added to a matrix made up of the x and y coordinate adjustments to give the new coordinates. Try plotting them to demonstrate that it has worked correctly. This transformation can be written in a more general format as:

$$\begin{bmatrix} x_1 & y_1 \\ x_2 & y_2 \\ . & . \\ x_n & y_n \end{bmatrix} + \begin{bmatrix} x_a & y_a \\ x_a & y_a \\ . & . \\ x_a & y_a \end{bmatrix}$$

where the first matrix contains the original coordinates and the second matrix contains the x and y adjustments.

It can also be demonstrated that a simple matrix multiplication will enable an image to be rescaled. If we wished to double the size of the square this can be achieved by multiplying the original coordinate matrix by a two by two diagonal matrix. This is shown below.

$$\begin{bmatrix} 1 & 1 \\ 1 & 3 \\ 3 & 3 \\ 3 & 1 \end{bmatrix} \cdot \begin{bmatrix} 2 & 0 \\ 0 & 2 \end{bmatrix} = \begin{bmatrix} 2 & 2 \\ 2 & 6 \\ 6 & 6 \\ 6 & 2 \end{bmatrix}$$

The x and y axes can be scaled differently. This will change the size and shape of the image. For example, if the horizontal dimensions of the square are tripled while doubling the y axis, a rectangle will be produced. This is shown below.

$$\begin{bmatrix} 1 & 1 \\ 1 & 3 \\ 3 & 3 \\ 3 & 1 \end{bmatrix} \cdot \begin{bmatrix} 3 & 0 \\ 0 & 2 \end{bmatrix} = \begin{bmatrix} 3 & 2 \\ 3 & 6 \\ 9 & 6 \\ 9 & 2 \end{bmatrix}$$

The general scale and shape transforming matrix can be represented by:

$$\begin{bmatrix} S_x & 0 \\ 0 & S_y \end{bmatrix}$$

where S_x and S_y are the scaling values for the x and y axes respectively.

If you have plotted the various coordinates you will have noted that the scale and shape conversions also move the origin of the shape. This could be corrected by a simple matrix addition (movements to the left and downwards are possible if negative values are used).

Shape rotations can also be achieved by the application of a little trigonometry combined with another matrix multiplication. The proof of this statement is outside the scope of this text but is given by Foley and Van Dam.

If a three dimensional image is required an additional z coordinate must be defined for each point. A cube can be represented by the six coordinate triplets that correspond to the six corners of the cube. These coordinates can then treated as a three column matrix. (In fact a four column matrix is used. The fourth column is used to hold an additional coordinate. This is because points are best represented by a system of homogeneous coordinates. This system was developed in geometry and any point in an N dimensional space is referenced by coordinates in $N + 1$ dimensions. In most circumstances this fourth coordinate will be 1 and can be effectively ignored although it must be included in coordinate matrices.) Since the screen has only two dimensions a three dimensional image can only be represented by its projection onto a two dimensional surface. There are a set of standard mathematical procedures which can be used to obtain the projection coefficients. These are beyond the scope of this book.

There are a number of other standard procedures which can be used to manipulate the coordinates of a three dimensional structure. For example, there are transformations available which will move, rotate and scale the structure. Others can be used to give perspective to a three dimensional image and also produce left and right stereoscopic images. If the stereoscopic images are printed in suitable colours and viewed through an appropriate pair of lenses the image will appear to possess three dimensions. This is a very useful technique for providing structural information about proteins and similar complex molecules and structures. By producing suitable rotations and stereoscopic images the molecule can be inspected from many directions.

The mathematics of these transformations is outside the scope of this book and any interested reader should again consult Foley and Van Dam.

Many of the more recent dialects of BASIC contain a variety of quite complex graphics commands. These will be absent from most mainframe dialects because they have been designed for use with teletypes which have very restricted display facilities. Table 9.2 lists five of the more complex graphics commands provided by Microsoft Extended Color BASIC. Other dialects may provide even more powerful graphics commands.

Table 9.2 More complex graphics commands

Command	Action and notes
CIRCLE(X,Y),R,options	As the name suggests this command will draw a circle whose centre is at the screen coordinates X,Y. The radius is set by R. Options include specifying a height to width ratio. This will produce an ellipse. It is also possible to set the start and end coordinates of an arc so that the circle is not completed.
LINE(X,Y)–(X,Y),options	This command is used to draw a line between a pair of start and end coordinates. The options include various colour options and the construction of a box whose baseline is defined by the pair of X,Y coordinates.
GET(X,Y)–(X,Y) and PUT(X,Y)–(X,Y)	This pair of commands works together. The GET command saves the information from a block of the display in an array. This information can then be PUT into a new screen position.
DRAW "string"	A string of cursor movement and plotting commands is built up from a collection of commands and stored as a string such as D$. The commands allow the cursor to be moved in eight directions: left, right, up, down and diagonally. A reasonably complex image can be established which can be drawn by the simple command DRAW "string". Additional facilities allow the finished drawing to be scaled and rotated.

Although modern dialects of BASIC provide a number of powerful graphics commands most of them are quite slow in operation and are generally restricted to two dimensional images. BASIC is the not the best language for advanced graphics. A compiled language such as Pascal or FORTRAN would be much more useful, particularly when combined with existing graphics software packages such as GINO-F (Graphical INput/ Output – FORTRAN). This a library of over 160 graphics subroutines that can used with a suitable display device such as a Tektronix T4010 terminal.

9.4 Biological applications

A cross-section of four biological applications of computer graphics and image processing techniques is introduced in this section. The mathematical algorithms which form an essential component of these examples will not be discussed in detail. A more detailed account of several biological applications can be found in *Computing in Biological Science* edited by Geisow and Barrett (1983). Also included at the end of this section is a simple graph plotting routine.

Image processing techniques can be used to extract information from photographs or television pictures. Image processing systems use either digital cameras, cameras whose output can be used directly by a computer, or video digitising hardware and software. A video digitiser can convert the output from a video camera into a digital format which can be processed by a computer. Both of these systems break an image down to a large number of pixels which are described individually by their score on a grey scale. A grey scale is a means of splitting the entire range of shades between white and black into a fixed number of elements. The amount of information which can be communicated by a grey scale depends upon the the number of incremental steps between white and black. As the number of steps increases so does the amount of memory required to hold the information. A high resolution image of 512 by 512 pixels which is described by a 64 element grey scale would require about a quarter of a million bytes of memory. This amount of information can only be held on hard disks or professional multitrack tape recorders. Complex computer image analysis techniques are generally beyond the capabilties of microcomputers.

9.4.1 Remote sensing

Aerial and satellite photographs can provide a detailed and comprehensive database of current land uses. In addition they can be used to monitor such diverse activities as forest productivity, pest and disease outbreaks, pollution damage and rates of vegetational change. These tasks can be simplified and quantified if the photographs can be subjected to computer analyses. This is the science of remote sensing.

There are two large databases which hold mainly satellite information. This information can be retrieved online (see Chapter 12) by other computers. The European Space Agency has a database file called LEDA which contains satellite information about Europe. The largest satellite database is EROS (Earth Resources Observation System). This database contains information from over 6,000,000 photographs. Most of the information is not, however, held as photographs. This is because the majority of the information has been provided by unmanned satellites (mainly the Landsat series). The satellites are equipped with a wide range of cameras and other sensing equipment which 'photograph' the earth using wavelengths not perceived by the eye. The information is transmitted to earth in a digital format which can then be used to produce photographs and displays. Because the information is effectively on a grey scale photographs would normally be black and white. It is, however, possible to split up this scale and produce colour photographs. These colour photographs are in fact 'false colour composites', in that the colours do not match the natural ones.

Vegetation usually appears to be red, while clear water is black! It is, of course, possible to produce the image in any combination of colours if the software is suitably adjusted.

The Institute of Terrestrial Ecology (ITE) in the United Kingdom is currently using satellite information, mainly Landsat, to investigate land usage and management in the UK. Each Landsat scene that they investigate covers an area of 185 by 185 km which requires 25 Mb of storage. Obviously this volume of information can be analysed efficiently only by large and expensive computer systems. (However, the ITE has been experimenting with an Apple IIe computer, equipped with an additional vector graphics processor and hard disk, as a cheap method for analysing partial Landsat images.)

9.4.2 Cytological investigations

Image analysis techniques are now becoming available as part of standard laboratory equipment. An example of this is the Leitz automated cytological investigation system, based on a computer controlled microscope. The computer can move the microscope stage very accurately and hence return to particular fields of interest with great precision. Focusing is also under computer control. This is achieved by the image analysis software which controls the movement of a stepping motor which in turn drives the focusing mechanism. The images from the microscope slide are recorded by a television camera whose output can be analysed and digitised by a minicomputer. This information can be used to produce a television picture which has a resolution of 256 by 256 pixels. More importantly the digitised information can be analysed for the occurrence of particular cytological features, for example, cancerous cells stain more deeply than normal cells when appropriate stains such as acriflavine–Feulgen are used. The software can focus onto subjects by maximising the contrast between the subject and its background. These subjects are then categorised by a transformation which relates the range of grey values within the subject to its area. The transformation produces a profile which can be compared with those expected for a number of potential subjects. The grey scale profile of a cancerous cell will be markedly different from that of normal cells and artifacts such as specks of dirt and clumps of stain. The stage coordinates of positive subjects will be recorded for later display to the operator. This allows the operator to check for false positives which may occur if, for example, two normal nucleii overlap on the slide.

9.4.3 Chromosome organisation

The next example combines image processing techniques with those of computer graphics. The work was reported by Mathog and his co-workers in the journal *Nature* (Vol 308, No 5958, pp 414–421, 1984).

It is becoming increasingly recognised that the spatial geometry of interphase chromosomes may be very important with respect to gene function and chromosome organisation. Particular chromosome orientations would bring into close proximity distant parts of the same chromosome and segments of non-homologous chromosomes. If these orientations are fixed the sequence of genes on a chromosome may not be sufficient to explain position effects. However, because of the nature of interphase nucleii these relationships are very difficult to establish. Normal microscopic methods produce photographs that are very difficult to analyse. Only by applying image analysis techniques can the complex relationships be simplified. The techniques developed by Mathog and his co-workers allowed them to analyse the three dimensional orientations of interphase chromosomes and then produce three dimensional models of the structures.

They investigated the chromosome organisation in cells from the salivary glands of Drosophila melanogaster. The chromosomes were stained with a fluorescent dye and a computer controlled microscope was used to produce images from a vertical stack of 24 focus planes through a nucleus. These images were digitised from a video camera to give a resolution of 512 by 512 pixels for each focus plane. This degree of resolution was sufficient to allow the identification of chromosome bands and hence the five major chromosomes could be localised within the intact nucleus. Since the position of each of the chromosome arms was now known for 24 cross-sections of the nucleus, a three dimensional model of their positions could be produced (by a computer graphic technique). Once the model was available it was used to investigate the properties of possible patterns and trends in chromosome organisation. This type of investigation would be impossible without the aid of computer graphics and image processing techniques.

9.4.4 Molecular models

Computer studies have played a large part in our understanding of the forces which determine the structure and associated functioning of many complex biological macromolecules. As computer graphic techniques become cheaper and more available they will play an increasing role in the study of such molecules. If the structure of a molecule is known (from chemical and X-ray crystallography data) a three dimensional model of the molecule can be generated on a display. Using the mathematical techniques outlined in section 9.3, the model can be rotated and viewed from various directions. Many biologists would probably find such a pictorial representation much less daunting than a mathematical description of the same molecule. If the structure has not been determined but chemical data is available it may be possible to combine the graphics program with one which can predict molecular structure (see Geisow and Barrett) from chemical data and hence produce a model which can be compared with similar molecules.

9.4.5 A simple graph plotting routine

It is often helpful to have a graphical representation of the relationships between two sets of numbers, graphs being easier to interpret than tables of numbers. It would be useful, therefore, if data could be represented graphically by a microcomputer. The following program is a little clumsy because it attempts to be a type of 'lowest common denominator' in providing a routine which will work with most microcomputers. Note that it requires a memory mapped display which may not be present in terminals connected to mini- and mainframe computers. Often the display produced by a terminal is nothing more than a glorified paper teletype display. This program could be vastly improved by using suitable values for the variables and employing any special facilities offered by the type of computer being used. The variables used are described in REM statements.

```
10 REM      A SIMPLE GRAPH PLOTTING ROUTINE FOR USE WITH
11 REM      A COMPUTER WHICH HAS A MEMORY MAPPED DISPLAY
12 REM      IT ASSUMES THAT COORDINATE 0,0 IS AT THE TOP LEFT
13 REM
14 REM      THE FOLLOWING VARIABLES ARE USED IN THE PROGRAM
15 REM       X(I) & Y(I) ARE THE ARRAYS OF RAW DATA
16 REM       XM & YM ARE THE LARGEST VALUES OF X & Y
17 REM       THESE MUST BE IDENTIFIED BEFORE THE ROUTINE CAN BE USED
18 REM       YO IS THE SCREEN COORDINATE FOR THE ORIGIN OF THE Y AXIS
19 REM       YN IS THE SCREEN COORDINATE FOR THE TOP OF THE Y AXIS
20 REM       XO & XN ARE EQUIVALENT POINTS FOR THE X AXIS
21 REM       SUITABLE VALUES FOR THE APPLE LOW RESOLUTION SCREEN ARE:
22 REM       YN = 5, YO = 35, XN = 38, XO = 5 (A 40 BY 48 BLOCK DISPLAY )
23 REM       XX(I) & YY(I) ARE ARRAYS CONTAINING THE SCREEN COORDINATES
24 REM       FOR THE X AND Y VALUES
25 REM       XL & YL ARE THE LENGTHS, IN SCREEN COORDINATES, OF THE AXES
26 REM       THE FOLLOWING PROGRAM USES APPLESOFT COMMANDS
27 REM       READ IN SOME TEST DATA
30  FOR I = 1 TO 10
31  READ X(I), Y(I)
32  NEXT I
39 REM       INITIALISE THE VARIABLES
40  LET XM = 95
41  LET YM = 88
42  LET YN = 5
43  LET YO = 35
44  LET XN = 38
45  LET XO = 5
46 REM  CALCULATE THE X AND Y AXIS LENGTHS. THIS IS USED TO SCALE THE
         X AND Y VALUES
47  LET XL = XN - XO
48  LET YL = (YO - YN ) * -1
49 REM  THE MULTIPLICATION BY -1 COMPENSATES FOR THE FACT THAT THE
         Y AXIS SCREEN COORDINATES ARE DECREASING WHILE THE ACTUAL
         VALUES OF Y ARE INCREASING
50 REM       SCALE THE X AND Y ARRAYS TO THEIR SCREEN COORDINATES
51  FOR I = 1 TO 10
52  LET XX(I) = INT (X(I) / XM * XL) + XO
53  LET YY(I) = INT (Y(I) / YM * YL) + YO
54  NEXT I
59 REM       PLOT THE GRAPH, START BY CLEARING THE SCREEN AND SELECTING
             THE LOW RESOLUTION SCREEN AND THE PLOTTING COLOUR
60  HOME
61  GR
62  COLOR = 15
63  HLIN 5,38 AT 35
64  VLIN 5,35 AT 5
```

```
65 REM   THESE TWO COMMANDS ARE USED TO DRAW THE AXES AT THE CORRECT PLACES
66  FOR I = 1 TO 10
67  PLOT XX(I), YY(I)
68  NEXT I
69 REM          HOLD THE DISPLAY IN PLACE
70  GOTO 70
100 DATA 12,4,24,15,36,26,50,43,60,54,72,65,85,80,44,47,69,66,95,88
```

9.5 Summary

Computer graphics and image processing techniques have many existing biological applications but their potential uses are enormous. At the present moment the work is only feasible with minicomputers which have access to large disk stores. However, the next generation of 32-bit microcomputers combined with the price reduction in 'Winchester' type hard disks should make computer graphics systems more freely available. These future and existing systems will also be very dependent upon software. There are many graphics software packages available for microcomputers but most of these are too slow and restricted to be of great value. It is quite probable that more powerful microcomputer software packages will become available in the next few years.

Chapter 10 Structured Programming

10.1 Historical background

Until quite recently a good programmer was identified as someone who could write clever and often devious programs which occupied the minimum quantity of main memory and executed in the fastest possible time. These qualities were important because of the restricted capabilities offered by the earlier computers. Unfortunately the programs that inevitably resulted from these equipment constraints were often exceedingly difficult to follow. They became even more difficult to understand as they were modified to correct bugs or to incorporate additional features.

Present day computers do not impose the same speed and memory constraints. Consequently the requirement for 'clever' programming has diminished. The current emphasis is on well structured programs which can be easily read, maintained and modified by other programmers. Languages such as Pascal were developed specifically to support these aims. BASIC was not developed with these criteria in mind. Although many recent dialects of BASIC support some of the important Pascal features it is still very easy to write bad BASIC programs. However, it is possible to write well structured programs in BASIC if sufficient care is taken over the process of program development and design. This is true even for the majority of BASIC dialects which do not include the Pascal-type commands.

10.2 What is structured programming?

Many people seem to define a structured program as being one which contains no GOTOs! There is, obviously, more to it than that. Structured programming should not be thought of as just a set of hard and fast programming rules but rather a style or attitude to programming. A structured program is likely to be one which was thoroughly thought out before the coding started. It will normally be composed of several blocks or modules. Each of these blocks should normally have:

   a maximum length of one hundred lines
   one entry point
   one exit point
   no redundant statements (i.e. never executed)
   no infinite loops.

In addition the blocks will normally contain code which falls into one of three categories of construction. These are:

sequence — one instruction is followed by the next, etc.;
selection — an instruction is followed by one of two alternatives;
repetition — an instruction which is repeated a number of times.

Wherever possible the algorithm and coding used will be presented in a manner which will maximise the readability of the program. There are several practices which fulfil this aim. These are:

a) keeping the logic as simple as possible even if it means sacrificing speed for readability;
b) using indentation of the program to highlight the logic used (this is not possible with all dialects);
c) careful selection of the variable names (use long descriptive names if available);
d) avoiding unnecessary GOTO statements;
e) using conditional loops wherever possible.

10.3 Structured commands in BASIC

In addition to the IF ... THEN and IF ... THEN ... ELSE constructions which were covered in Chapter 6 there are two other structured commands available in some of the recent BASIC dialects. These are the REPEAT ... UNTIL and WHILE ... WEND constructions. Both of these constructions are used to control the execution of conditional loops. The REPEAT ... UNTIL construction allows for the execution of a set of commands until a specified condition is true. Conversely, the WHILE construction facilitates the execution of a set of commands as long as a specified condition is true. Although they are very similar commands they have an important difference which relates to when the specified condition is tested. The statements between the REPEAT and UNTIL commands will always be executed at least once since the condition is not tested until the end of the construction. In the case of the WHILE construction, the specified condition is tested prior to the execution of any statements controlled by this construction. Consequently, they will be skipped if the condition is false. These differences are illustrated below.

General command format
REPEAT ... UNTIL   *WHILE*
REPEAT   WHILE condition is true
--- statements which are to be executed ---
UNTIL condition is true   WEND (While END)

In the following examples both constructions will be used in short programs to sum a list of numbers of indeterminate length. The end of the list will be identified by entering 0 as one of the numbers to be summed. In these examples indentation of statements and long variable names will be used even though they are not universally available.

116

```
10 REM --REPEAT UNTIL DEMONSTRATION--    10 REM--WHILE DEMONSTRATION--------
11 REM  SUM = TOTAL, NUMBER = NUMBER     11 REM  SUM = TOTAL, NUMBER = NUMBER
12 REM  COUNTER = COUNTER                12 REM  COUNTER = COUNTER
20 REM-------INITIALISATION----------    20 REM-------INITIALISATION--------
21 LET SUM = 0                           21 LET SUM = 0
22 LET NUMBER = 0                        22 LET NUMBER = 1
23 LET COUNTER = 0                       23 LET COUNTER = 0
25 PRINT "SUM AND AVERAGE"               25 PRINT "SUM AND AVERAGE"
26 PRINT                                 26 PRINT
30 REM-------CONDITIONAL LOOP--------    30 REM-------CONDITIONAL LOOP------
31 PRINT "ENTER NUMBER , 0 TO END"       31 PRINT "ENTER NUMBER , 0 TO END"
32    REPEAT                             32    WHILE NUMBER <> 0
33       INPUT NUMBER                    33       INPUT NUMBER
34       LET SUM = SUM + NUMBER          34       LET SUM = SUM + NUMBER
35       LET COUNTER = COUNTER + 1       35       LET COUNTER = COUNTER + 1
36    UNTIL NUMBER = 0                   36    WEND
40 LET COUNTER = COUNTER - 1            40 LET COUNTER = COUNTER - 1
50 REM-------USER SUMMARY-----------    50 REM-------USER SUMMARY----------
51 PRINT "TOTAL IS "; SUM               51 PRINT "TOTAL IS "; SUM
52 PRINT "NUMBER OF ITEMS "; COUNTER    52 PRINT "NUMBER OF ITEMS "; COUNTER
53 PRINT "MEAN IS "; TOTAL / COUNTER    53 PRINT "MEAN IS "; TOTAL / COUNTER
60 END                                  60 END
```

The two programs have only one difference, except for the type of conditional loop. This difference can be seen on line number 22. If line 22 in the WHILE program had initialised NUMBER to 0 the loop would never have been executed since the condition on line 32 would not be satisfied. Note also that TOTAL was not used as a variable name. This is because most dialects, even those which allow long variable names, will not accept names which contain BASIC statements. TOTAL contains the BASIC command TO.

In both programs the value of COUNTER is incremented after 0, which identifies the end of the list, has been entered. (0 is also added to the total but this does not affect its true value.) Line 40 has been included to reduce COUNTER to its correct value. Line 35 could have been written as:

```
35 IF NUMBER < > 0 THEN COUNTER = COUNTER + 1
```

Since this complicates the logic and would also produce an unnecessary delay in execution of the program it has not been used.

10.4 Writing structured programs

One of the best methods of developing a structured program is to use the 'top down' approach which begins by identifying the purpose of the program. Once this has been identified an algorithm can be developed. This begins by breaking down the problem into a small number of major conceptual blocks which will be identified by labels such as 'Read in data', 'Transform data'. If necessary, each of these major blocks is broken down into smaller blocks. Only the small program blocks are coded. This is shown diagrammatically in Fig. 10.1.

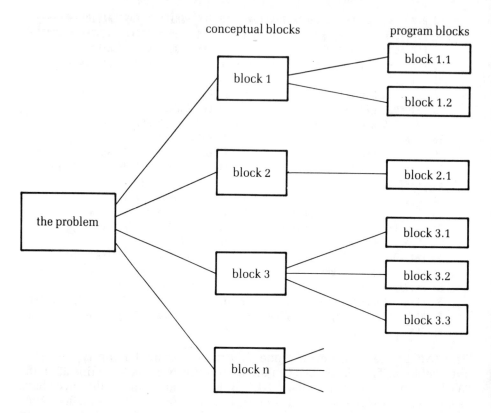

conceptual blocks   program blocks

Fig. 10.1 The 'top down' approach to structured programming

These smaller blocks can be initially coded in a type of language called pseudocode. This is not a programming language but an intermediate between English and many of the high level languages. A program which is developed in this way can be subsequently coded into a number of different languages. Each of these blocks may be thought of as separate programs. Ideally they should be written and tested independently of each other and then stored on disk or tape. They will only be assembled to form the main program quite late in the development phase. Unfortunately this is not always very easy to achieve on many microcomputers since it involves the merging (or appending) of programs. It can be a very useful process since the same blocks can often be used with either no, or very little modification, in many different programs.

The structure of most BASIC dialects does not simplify the task of writing and assembling subprograms and subroutines. There are three main problems to be overcome.

a) Since BASIC is a line orientated language particular care must be taken over the numbering of subprograms to avoid any clash of line numbers when the blocks are eventually assembled to form the complete program. Some computers do have a renumbering facility available but this does not overcome the fundamental problem.

118

b) A related problem is that, in most dialects, subroutines must be called by line number rather than by name. A statement such as GOSUB 1000 is much less meaningful than GOSUB DATAINPUT. A number of the more recent dialects have adopted a Pascal-type structure called the procedure (PROC). The procedure is defined in a similar manner to the user defined function. The start of the procedure is marked by the command DEF PROC Procedurename and terminated by ENDPROC. A procedure which has been defined in this way can be called by PROC Procedurename.

c) In most dialects of BASIC any variable names used are global. This means that whenever a particular variable name is used it will always refer to the same quantity. This makes it difficult to build up a library of subroutines or programs since there are likely to be name clashes between programs. Languages such as Pascal and FORTRAN, which make frequent use of subroutine calls, avoid this problem by allowing variables to be classed as global or local. A local variable is one whose influence is restricted to the subroutine or procedure in which it is defined. It is quite possible to have local and global variables with the same name. They will not interfere with each other. The local variables may be thought of as operating in a similar manner to the dummy variables in user defined functions. Values can be passed to and from these local variables in the form of global variables.

10.5 Example

In this example a program, which can be used to sort a list of numbers into ascending order, will be developed. Sorting routines form an important part of many programs. A biologist is quite likely to encounter them in programs written to perform the nonparametric statistic tests which form such an important part of many behaviour experiments. There is a large body of programming theory which is directed at developing efficient sort algorithms. The algorithm used in this program is one of the simplest but less efficient examples. The most efficient algorithms are based on quite complex logic which is not appropriate for this example. Efficiency is measured as the time needed to sort a list of objects.

The algorithm used here is known as the 'bubble sort'. It is given this name because the smallest numbers 'float' to the top of the list. Consider the following short list of numbers and the method used to sort them.

5
7
9
3
1

One method is to look down the list and compare adjoining pairs of numbers. If the second of the pair is smaller than the first, the two are swapped. The largest number will 'sink' to the bottom after one pass through the list, but the smallest numbers may not have made it up to the top. Consequently the process is repeated. The entire list need not be reconsidered since the largest number will be at the bottom and the unsorted list will be one number less in length. On the second pass the second largest number will 'sink' down to its correct position in the list. Thus the unsorted list is again reduced in length. This process is repeated until the unsorted list is one number long. Since this must be the smallest number the sort is now complete. This operation is shown below.

Pass number 0 1 2 3 4

		5	5	5	3	1
List		7	7	3	1	3
sequence		9	3	1	5	5
		3	1	7	7	7
		1	9	9	9	9

This algorithm can be expressed in psuedocode as follows.

> Obtain the length of the list to be sorted.
> Set the unsorted list length equal to the full list length.
> Initialise a counter to one.
> While the unsorted list is longer than one element in length:
>> While the counter is less than the number of items in the unsorted list:
>>> If a number is less than that which follows:
>>> Swap these two values.
>> Add one to the counter.
>> Endwhile.
>> Set counter back to one.
>> Decrease unsorted list length by one.
> Endwhile.

Note that even if your computer does not allow the use of conditional loops or indentation the pseudocode can be written in the above style. This style of internal or external documentation is a useful way of clarifying the logic used in an algorithm.

In the first example a BASIC program will be written which uses the WHILE ... WEND construction. This will be followed by an example which uses standard BASIC statements to mimic conditional loops.

```
10 REM    BUBBLE SORT ROUTINE ( SMALLEST FIRST ) VERSION 1
11 REM    VARIABLES USED :
12 REM    NUMBER( ) - THE ARRAY OF NUMBERS TO BE SORTED
13 REM    SORTED    - NUMBER OF ITEMS IN THE ARRAY
14 REM    UNSORTED  - LENGTH OF THE STILL UNSORTED LIST
15 REM    COUNTER   - COUNTER TO MARK THE CURRENT POSITION IN THE LIST
16 REM    TEMPORARY - A TEMPORARY VARIABLE USED DURING EXCHANGE OF VALUES
20 REM    ----------------------INITIALISATION----------------------------
21 REM    DIMENSION NUMBER( ) FOR LISTS LONGER THAN 10 ITEMS
22 LET SORTED = 0
23 LET UNSORTED = 0
```

```
24 LET COUNTER = 0
25 LET TEMPORARY = 0
30 REM -----------READ IN AND PRINT OUT THE ITEMS TO BE SORTED------------
31 REM            999 MARKS THE END OF THE LIST
32 PRINT "ORIGINAL UNSORTED LIST"
33 WHILE NUMBER( COUNTER ) <> 999
34     COUNTER = COUNTER + 1
35     READ NUMBER( COUNTER )
36     IF NUMBER( COUNTER ) <> 999 THEN PRINT NUMBER( COUNTER )
37 WEND
38 LET SORTED = COUNTER - 1
39 LET UNSORTED = SORTED
40 REM ----------------------SORT ROUTINE--------------------------------
41 LET COUNTER = 1
42 WHILE UNSORTED > 1
43     WHILE COUNTER < UNSORTED
44         WHILE NUMBER( COUNTER ) > NUMBER( COUNTER + 1 )
45             LET TEMPORARY = NUMBER ( COUNTER )
46             LET NUMBER( COUNTER ) = NUMBER ( COUNTER + 1 )
47             LET NUMBER( COUNTER + 1 ) = TEMPORARY
48         WEND
49         LET COUNTER = COUNTER + 1
50     WEND
51     LET COUNTER = 1
52     LET UNSORTED = UNSORTED - 1
53 WEND
60 REM ----------PRINTOUT OF LIST TO DEMONSTRATE THAT METHOD WORKS--------
61 PRINT "SORTED LIST"
62 LET COUNTER = 1
63 REPEAT
64     PRINT NUMBER( COUNTER )
65     LET COUNTER = COUNTER + 1
66 UNTIL COUNTER = SORTED
67 REM -----DATA FOR SORTING(LESS THAN 10 ITEMS IF NUMBER( ) IS UNDIMENSIONED)
68 DATA 5,7,9,3,1,999
70 END
```

The program follows the pseudocode quite closely. After the data has
been read in by the lines 33 to 37 the length of the list is set as being
equal to 1 less than the current counter value. This is to remove the effect
of the unwanted data item 999. A conditional statement was included on
line 36 to ensure that the unwanted 999 was not printed as part of the
unsorted list. The sorted list is printed by using a REPEAT ... UNTIL
construction on lines 63 to 65. The exchange process, which occupies
lines 45 to 47, can be replaced by the very useful SWAP command in
some dialects. As the name suggests it is used to swap the values of two
variables, e.g. SWAP NUMBER(COUNTER),NUMBER(COUNTER+1). Line
number increments of 1 have been used in most parts of the program.
This leaves no room for additional code (unless the program is renum-
bered). If a program is well thought out this should not matter since
nothing should have been forgotten!

The second example is written in a dialect which allows only the simplest variable names and only one conditional construction, IF ... THEN. Nevertheless, it follows closely the logic contained in the pseudocode.

```
10 REM       BUBBLE SORT ROUTINE ( SMALLEST FIRST ) VERSION 2
11 REM       VARIABLES USED:
12 REM       N( ) - THE ARRAY OF NUMBERS TO BE SORTED
13 REM       L1   - NUMBER OF ITEMS IN THE ARRAY
14 REM       L2   - LENGTH OF THE STILL UNSORTED ARRAY
15 REM       C    - COUNTER TO MARK THE POSITION IN THE LIST
16 REM       T    - TEMPORARY VARIABLE USED DURING EXCHANGE OF VALUES
20 REM ----------------- INITIALISATION --------------------------
21 REM       DIMENSION N( ) FOR LISTS LONGER THAN 10 ITEMS
22 LET L1 = 0
23 LET L2 = 0
24 LET C = 0
25 LET T = 0
28 REM ------- READ IN DATA AND PRINT OUT LIST TO BE SORTED ---------
29 PRINT "ORIGINAL UNSORTED LIST"
30 REM WHILE NUMBER( COUNTER ) <> 999
31 IF N(C) = 999 THEN GOTO 36
32    LET C = C + 1
33    READ N(C)
34    IF N(C) <> 999 THEN PRINT N(C)
35    GOTO 31
36 REM WEND
37 LET C = C - 1
38 LET L1 = C
39 LET L2 = L1
40 REM ------------------ SORT ROUTINE ----------------------------
41 LET C = 1
42 REM WHILE UNSORTED > 1
43 IF L2 = 1 THEN GOTO 60
44    REM WHILE COUNTER < UNSORTED
45    IF C >= L2 THEN GOTO 55
46        REM WHILE NUMBER( COUNTER ) > NUMBER( COUNTER + 1 )
47        IF N(C) < N(C+1) THEN GOTO 51
48            LET T = N(C)
49            LET N(C) = N(C+1)
50            LET N(C+1) = T
51        REM WEND
52        LET C = C + 1
53    REM WEND
54    GOTO 44
55    LET C = 1
56    LET L2 = L2 - 1
57 REM WEND
58 GOTO 42
60 REM ------- PRINTOUT TO DEMONSTRATE THAT THE METHOD WORKS ---------
61 PRINT "SORTED LIST"
62 LET C = 1
```

```
63 REM REPEAT
64    PRINT N(C)
65    LET C = C + 1
66 REM UNTIL COUNTER = SORTED
67 IF C <= L1 THEN GOTO 63
68 REM --------- DATA FOR SORTING, MARK END WITH 999 ----------------
69 DATA 5,7,9,3,1,999
70 END
```

REM statements have been used to identify the start and end points of the WHILE and REPEAT loops and the equivalent conditions are tested by IF ... THEN statements. GOTOs are used to force reruns of the loops. The use of GOTO in this type of construction is quite legitimate, even within the constraints of structured programming.

10.6 Summary

It may seem to you that structured programming involves a lot of work and produces much longer programs than are absolutely necessary. To a certain extent both of these observations are true. However, you should remember that the type of problems which can be dealt with at this stage of your programming experience are relatively simple. As the problems become more complex the benefits of a structured approach will become more tangible. It should be remembered that a good program is one which always works correctly. Furthermore, there is no credit to be gained from developing an algorithm whose logic is so tortuous that no one, including yourself, can understand how it works. A good program is ,therefore, one which not only works correctly but can be seen to do so. This result is most likely to be obtained if the 'rules' of structured programming are observed.

10.7 Problems

1. How could the sort algorithm be made more efficient? (Hint — suppose a list of 10 numbers was in the correct sequence after one pass.)

2. Write an algorithm, in pseudocode, which could be used to search through a datafile to count and identify those records in which the value of a variable, YIELD( ), was greater than 150.

3. Code the algorithm developed in problem 2 using REPEAT ... UNTIL or WHILE ... WEND constructions and then using IF ... THENs to replace the structured commands.

Chapter 11 Computer Models

11.1 Introduction

The concept of models usually brings forth images of physical models, i.e. small scale three dimensional structures. Although physical models can be a very useful means of describing the structure and function of biological systems, such as organs, other types of model are in regular, and occasionally unrecognised, use. A model is a representation of a system or process which will always include a degree of simplification in comparision to the original. Models are important tools in both biological research and teaching. Three main types of computer model can be recognised, the first of which uses computer graphics to represent biological structures ranging from molecules to diagrammatic representations of large biological systems. Computer graphics was covered in Chapter 9 and will not, therefore, be considered here. The two remaining categories of computer models are interrelated: mathematical and simulation models.

A mathematical model is a numerical representation of a process or activity which is capable of undergoing change. Many biological processes can be modelled in this way. All students of biology will, at some time, have used equations which model a biological activity. The computer is useful because solutions can be rapidly obtained for even quite complex mathematical models. A computer simulation is an implementation of a mathematical model so that a process can be monitored over a period of time. It is easy to become overpowered by mathematical models, especially those which have been used with computers. It should be remembered that all models are an approximation of the original system. Consequently, the output from a model or simulation should always be compared with the behaviour of the real system. Models are only of value if they make biological as well as mathematical sense.

An excellent introduction to BASIC microcomputer models in biology can be found in Spain's book (see bibliography for details).

The process of computer modelling will be illustrated by developing programs for three simple biological models. The programs will be written using the 'rules' of structured programming, as described in Chapter 10. Each model will be introduced by a brief summary of the biological theory upon which it is based.

11.2 Model 1 – energy balance in living organisms

This model is based on equations described by Gates (1972). Additional information about this topic can be found in his book (see bibliography).

This model will be concerned with the physical laws which govern energy exchanges between an organism and its surroundings. Any surface with a temperature of greater than 0 K (−273°C) will lose energy in the form of electromagnetic radiation. The wavelength of this radiation is related to the surface temperature. At temperatures which are normally found on the earth's surface most of this emitted energy will be restricted to infra red wavelengths. The amount of radiation emitted by a surface can be found from Stefan-Boltzmann's law. This is a mathematical model which describes some of the energy exchange processes which occur at an organism's surface.

$$\text{radiation emitted} = \sigma\, T^4 \text{ J cm}^{-2}\text{ min}^{-1} \tag{11.1}$$

σ is the Stefan-Boltzmann constant which is 3.39834×10^{-10} J min^{-1} cm^{-2}K^{-4}

This law is only true for a perfect radiator. Organisms are not perfect radiators, they emit only a proportion of the expected energy. Equation 11.1 can be modified to take account of this.

$$\text{radiation emitted} = \epsilon \sigma\, T^4 \tag{11.2}$$

ϵ is the emissivity of the surface, for most organisms this is about .95

Organisms are constantly absorbing energy from a variety of sources. In the case of plants most of the absorbed energy is in the infra red region. If an organism is to survive it must achieve an equilibrium such that the amount of energy gained is equal to the amount lost. Obviously complete control is impossible and at any one time an organism may be in a positive or negative imbalance. Positive imbalance, i.e. a surfeit of energy, will be associated with a rise in temperature and vice versa. Major deviations from the balance point may be fatal. It is, therefore, very important that organisms have some means of controlling this relationship.

The simplest model of this energy exchange process assumes that an organism can only lose energy by radiation (as predicted by the Stefan-Boltzmann law). It is possible to predict what the surface temperature of an organism will be for any particular radiation regime. If the organism has reached equilibrium then

$$\text{radiation absorbed} = \text{radiation emitted}$$

Since radiation emitted is equal to $\epsilon \sigma\, T^4$ then at equilibrium radiation absorbed $= \epsilon \sigma\, T^4$. If the radiation absorbed is known the equation can be rearranged to solve for T (the surface temperature).

$$T = \sqrt[4]{\frac{R}{\epsilon\sigma}} \tag{11.3}$$

It will be a useful exercise for you to write a short program which can be used to predict the surface temperature of a leaf experiencing the following radiation regimes (where R is the amount of radiation in J cm^{-2}min^{-1}).

$R = 2.40$, this is equivalent to a clear winter day in Manchester, UK

$R = 6.00$, this is equivalent to a clear day in a hot desert

You will find that the leaf temperatures in the desert are too high for survival (remember that the temperature is in degrees Kelvin). Evidently the model is not very good! It can be improved by considering some additional physical and biological processes. Organisms have additional mechanisms for losing heat. The first one to consider is convection.

The rate at which energy is lost by convection is determined by:

1. the temperature difference between the surface of an organism and its surroundings

2. the convection coefficient of the organism (h_c).

If the simplest assumptions are used it can be shown that the convection coefficient of an organism is determined by the wind speed and the surface geometry. The convection coefficient is predicted by equations 11.4 and 11.5, which are themselves mathematical models.

Convection coefficient (h_c):

for a flat, horizontal plate (plant leaf)

$$h_c = 2.395 \times 10^{-2} \sqrt{\frac{\text{wind speed (cm}^{-1})}{\text{leaf width (cm)}}} \ \text{J cm}^{-2} \text{ min}^{-1} \ {}^\circ\text{C}^{-1} \quad (11.4)$$

for a cylinder (succulent leaf, pine needles, animals)

$$h_c = 2.579 \times 10^{-3} \sqrt{\frac{\text{wind speed}}{\text{(cylinder diameter in cm)}^2}} \quad (11.5)$$

The amount of energy lost by convection is predicted by equation 11.6, yet another model!

$$\text{energy lost by convection (J cm}^{-2} \text{ min}^{-1}) = h_c (T - T_a) \quad (11.6)$$

where T_a = medium temperature (air temperature)

This additional mechanism for energy loss can now be incorporated into the energy balance equation (11.3).

radiation absorbed = radiation emitted + convective heat loss

$$R = \epsilon \sigma T^4 + h_c (T - T_a) \quad (\text{J cm}^{-2} \text{ min}^{-1}) \quad (11.7)$$

This is a more useful model because it can be used to investigate how a leaf temperature would vary with changes in size and shape, changes in wind speed and different radiation regimes. Some of the differences in plant morphology between desert and non-desert species can be related to environmental conditions by using predictions from this model. Note how the model becomes more realistic as it becomes more complex!

Unfortunately solving this equation for T when all other parameters are known is not a simple process. Rearrangement of equation 11.7 gives a quartic equation (one which has a term raised to the power 4). Unlike quadratic equations, there are no easy solutions to quartic equations.

$$R + h_c T_a = \epsilon \, \sigma \, T^4 + h_c T \qquad (11.8)$$

Since all quantities except T are known, equation 11.8 can be simplified to:

$$x = yT^4 + zT \qquad (11.9)$$

The simplest way of determining the real, and biologically meaningful, value of T which satisfies equation 11.9 is by iteration. Iteration is a process which begins by guessing a value for an unknown quantity which is used to solve the equation. The resulting solution, or estimate, is compared with the correct solution. If the estimated solution is too large the guess must have been too large and it should be reduced. If the estimated solution becomes too small then the guess must be increased. The correct value of the unknown is gradually approached by continually reducing the magnitude of change in the guess. It is usually necessary to specify a degree of precision, i.e. a minimum value is set for the difference between the true solution and that produced by the estimate. At this point the iteration will cease. In equation 11.9 we require a value for T, the leaf temperature. All other variables are known. The iteration should be started with a guess which is at least in the correct order of magnitude. The leaf temperature should be reasonably similar to the air temperature, therefore the iteration can begin by setting the leaf temperature equal to the air temperature. This value is then used to solve the equation for the value of x (whose true value is known). The estimated solution will differ from this true value and the leaf temperature can be raised or lowered to take account of this. The equation is again solved, this time using the new value for the leaf temperature. Ideally, this process should be repeated until the estimate is the same as the true value. In practice, the iteration is stopped when the difference between the two solutions is negligible. In the programs which follow iteration ceases when the difference between the estimated and true solutions is less than 0.1.

The following programs contain translations of the mathematical model of energy exchange, described by equation 11.7, into BASIC statements. Example 11.1 uses long variable names and a WHILE ... WEND loop is used to control the iteration of the leaf temperature. An equivalent program, Example 11.2, which uses the minimum set of BASIC facilities is also presented.

Note
These programs use both very small and very large numbers which are usually represented in an exponential format. A number such as 0.0000323 can be written as 3.23E–05 and 84520000000 as 8.452E+10. These are directly equivalent to 3.23×10^{-5} and 8.452×10^{10} respectively. This type of number representation is explained in more detail in Appendix A.

Example 11.1

```
10 REM         LEAF TEMPERATURE MODEL WITH STRUCTURED COMMANDS
11 REM     VARIABLES USED
12 REM         LD - LEAF DIAMETER, WINDSPEED - WINDSPEED
13 REM         R - ABSORBED RADIATION, AIRTEMP - AIR TEMPERATURE
14 REM         LEAFTEMP - LEAFTEMP, HC - CONVECTION COEFFICIENT
15 REM         SB - STEFAN BOLTZMANN CONSTANT x EMISSIVITY ( 0.95 )
16 REM         ESTIMATE - ESTIMATED SOLUTION, ACTUAL - REAL SOLUTION
17 REM         REFINEMENT - VALUE USED TO REFINE THE GUESS
18 REM         DIFFERENCE - DIFFERENCE BETWEEN ACTUAL AND ESTIMATED SOLNS
19 REM         PRECISION - THE LEVEL OF ACCURACY
20 REM -------------- INITIALISATION -----------------------------------
21 REM         CLEAR THE SCREEN
30 LET PRECISION = .1
50 PRINT "ENTER THE FOLLOWING QUANTITIES: "
60 PRINT "     LEAF DIAMETER IN CM ";
70 INPUT LD
80 PRINT "     WINDSPEED IN CM PER SECOND ( MINIMUM OF 10 ) ";
90 INPUT WINDSPEED
100 PRINT "     INCIDENT RADIATION ( J PER SQ CM PER MIN ) ";
110 INPUT R
120 PRINT "     AIR TEMPERATURE IN CELSIUS ";
130 INPUT AIRTEMP
132 REM A VALUE OF 0 WILL CAUSE COMPUTATIONAL PROBLEMS
134   IF AIRTEMP = 0 THEN AIRTEMP = .1
140 LET LEAFTEMP = AIRTEMP + 273
150 LET SB = 3.39834E-10 * 0.95
160 IF WINDSPEED < 10 THEN WINDSPEED = 10
170 LET HC = 2.395E-02 * SQR ( WINDSPEED / LD )
180 LET ESTIMATE = SB * LEAFTEMP^4 + HC * LEAFTEMP
190 LET ACTUAL = R + HC * ( AIRTEMP + 273 )
200 LET REFINEMENT = ABS ( AIRTEMP )
205 LET DIFFERENCE = ACTUAL - ESTIMATE
210 REM --------------- ITERATION LOOP -----------------------------------
220 WHILE ABS ( DIFFERENCE ) > PRECISION
230     IF ESTIMATE < ACTUAL THEN LEAFTEMP = LEAFTEMP + REFINEMENT
                               ELSE LEAFTEMP = LEAFTEMP - REFINEMENT
240     LET ESTIMATE = SB * LEAFTEMP^4 + HC * LEAFTEMP
245     LET DIFFERENCE = ACTUAL - ESTIMATE
250     LET REFINEMENT = REFINEMENT / 1.75
260     PRINT DIFFERENCE, LEAFTEMP
270 WEND
280 PRINT "LEAF TEMPERATURE AT ENERGY BALANCE WILL BE "; LEAFTEMP -273 ; " C"
290 END
```

Example 11.2

```
10 REM         LEAF TEMPERATURE MODEL - UNIVERSAL BASIC DIALECT
11 REM     VARIABLES USED
12 REM         D - LEAF DIAMETER, W - WINDSPEED
```

```
13 REM      R - ABSORBED RADIATION, T - AIR TEMPERATURE
14 REM      L - LEAF TEMPERATURE, H - CONVECTION COEFFICIENT
15 REM      S - STEFAN BOLTZMANN CONSTANT * EMISSIVITY (0.95)
16 REM      E - ESTIMATED SOLUTION, A1 - ACTUAL SOLUTION
17 REM      G - VALUE USED TO REFINE THE GUESS
18 REM      F - DIFFERENCE BETWEEN ACTUAL AND ESTIMATED SOLNS
19 REM      P - PRECISION OF THE ANSWER
20 REM --------------------- INITIALISATION -----------------------------
21 REM CLEAR THE SCREEN
30 LET P = .1
50 PRINT "ENTER THE FOLLOWING QUANTITIES: "
60 PRINT "   LEAF DIAMETER IN CM ";
70 INPUT D
80 PRINT "   WINDSPEED IN CM PER SECOND ( MINIMUM OF 10 ) ";
90 INPUT W
100 PRINT "   INCIDENT RADIATION ( J PER SQ CM PER MIN ) ";
110 INPUT R
120 PRINT "   AIR TEMPERATURE ( CENTIGRADE ) ";
130 INPUT T
132 REM A VALUE OF 0 WILL CAUSE COMPUTATIONAL PROBLEMS
134 IF T = 0 THEN T = .1
140 LET L = T + 273
150 LET S = 3.39834E-10 * 0.95
160 IF W < 10 THEN W = 10
170 LET H = 2.395E-02 * SQR ( W / D )
180 LET E = S * L^4 + H * L
190 LET A1 = R + H * ( T + 273 )
200 LET G = ABS ( T )
205 LET F = A1 - E
210 REM ---------- ITERATION LOOP WITH IMPLIED WHILE WEND ------------
220 IF ABS ( F ) < P THEN GOTO 280
230 IF E < A1 THEN L = L + G
235 IF E > A1 THEN L = L - G
240 LET E = S * L^4 + H * L
243 LET F = A1 - E
245 LET G = G / 1.75
260 PRINT F, L
270 REM WEND
271 GOTO 220
280 PRINT "LEAF TEMPERATURE AT ENERGY BALANCE WILL BE "; L - 273; " C "
290 END
```

A windspeed of 10 cm s^{-1} is assumed to be equivalent to a windless day. The value entered is checked on line 160 and adjusted upwards if it is less than 10. Line 134 is used to slightly alter the value of T if it has a value of zero. This is because subsequent multiplications by zero would cause problems. On line 140 the leaf temperature is initially set to the absolute air temperature (in degrees Kelvin). The absolute value is used to allow temperatures of less than zero to be used. This initial value for the leaf temperature is used on line 180 to produce an estimate of the actual quantity calculated on line 190. Finally, in the initialisation block, the value used to refine the estimate is set to the air temperature.

The iteration loop occupies lines 220 to 270 in Example 11.1 and lines 220 to 271 in Example 11.2. In both cases the iteration is continued as long as the absolute difference between the estimated and actual solutions is greater than 0.1. The absolute difference is used because it allows for both positive and negative differences. The first part of the iteration adjusts the value of the leaf temperature by adding or subtracting the value of the refinement. Following this, the estimated solution is recalculated. The refinement adjustment must now be reduced so that the difference between estimated and true solutions gradually decreases. A division by 1.75 produces reasonable results. Finally, the difference between the two solutions is printed along with the current leaf temperature value (degrees K). This step is not essential but it does illustrate how the iterative process works. When the two solutions are within 0.1 of each other the iteration ceases and the leaf temperature at energy balance is printed in degrees Celsius. If an unrealistic combination of variables is entered it is possible that the iteration will go into an infinite loop. Ideally the program should contain a section of code which tests for this eventuality.

Sample run

```
>RUN

ENTER THE FOLLOWING QUANTITIES
   LEAF DIAMETER IN CM ? 1
   WINDSPEED IN CM PER SECOND ( MINIMUM OF 10 ) ? 10
   INCIDENT RADIATION IN J PER SQ CM PER MIN ? 2.4
   AIR TEMPERATURE IN CELSIUS ? 30

 2.87898            273
 1.08591            290.143
 .0189705           299.939
LEAF TEMPERATURE AT ENERGY BALANCE WILL BE 26.9388 C

> OK
> RUN

ENTER THE FOLLOWING QUANTITIES:
   LEAF DIAMETER IN CM ? 1
   WINDSPEED IN CM PER SECOND ( MINIMUM OF 10 ) ? 100
   INCIDENT RADIATION IN J PER SQ CM PER MIN ? 6
   AIR TEMPERATURE IN CELSIUS ? 30

-5.15521            333
-.292778            315.857
 2.43395            306.061
 .880226            311.659
-.0128784           314.858
LEAF TEMPERATURE AT ENERGY BALANCE WILL BE 41.8575 C

> OK
```

130

Obviously this model is still a simplification of the energy exchange processes which are taking place at the surface of a leaf. Even so, the model produces values for leaf temperature which are close to those found in nature under the specified conditions. The model could be improved by taking into account the remaining major mechanism by which plants lose energy. This is of course water evaporation. The amount of energy lost by water evaporation (transpiration) is equal to the latent heat of evaporation for water (2429 J g^{-1} water) multiplied by the evaporation rate (g cm^{-2} min^{-1}). The rate of evaporation can be predicted (by yet another mathematical model) if certain leaf and physical characteristics are known.

$$\text{rate of evaporation} = \frac{d_1 + rhd_a}{r_1 - r_a} \qquad \text{g water cm}^{-2}\text{ min}^{-1} \qquad (11.10)$$

where d_1 and d_a are the saturation concentrations (g cm^{-3}) for leaf and air respectively

rh relative humidity (%)

r_1 internal resistance to water vapour diffusion from the leaf (min cm^{-1})

r_a resistance to water vapour diffusion across the boundary layer of still air that surrounds the leaf (min cm^{-1})

The full energy balance equation for a leaf can now be expressed as:

$$\text{energy in (incident radiation)} = \epsilon\, \sigma\, T^4 + \text{convection} + \text{transpiration} \qquad (11.11)$$

Although this information can be included in the leaf temperature model a potentially more valuable use for this model is the prediction of irrigation requirements. If the leaf temperature is known for a particular set of incident radiation, air temperature and wind speed regimes the model can be used to determine the rate of water evaporation that will satisfy the equation. This is a much simpler process since it does not require iteration.

11.3 Model 2 – selection against a recessive allele

The frequencies of alleles in a large, randomly mating population can be described by the Hardy-Weinberg law. This is a mathematical model which is often used by biologists. The frequencies of two allelic genes are represented by p and q such that $p + q = 1$. It is possible to derive an equilibrium from this relationship (assuming that a number of important assumptions are valid). The equilibrium genotype frequencies are:

Genotype	Equilibrium frequency
Homozygous dominant	p^2
Heterozygote	$2pq$
Homozygous recessive	q^2

These frequencies will remain constant as long as there is no force operating to change them. It is, therefore, possible to use this model as a basis for more complex models that investigate those forces which can change the allele frequencies within a population. If the genetic structure of a population changes, the population is said to be evolving. Mathematical models have been developed for many of the mechanisms which can produce changes in allele, and hence genotype, frequencies. This model is concerned with the effects of natural selection against a recessive allele.

The fitness of a phenotype (genotype) is a measure of the probability of reproductive success for an organism showing that phenotype. The difference in fitness between genotypes is expressed mathematically as a selection coefficient (s). The selection coefficient may have a value between 0 and 1. A value of zero is equivalent to no selection while a value of 1 indicates that the genotype is lethal (maximum selection). If 75 out of 100 organisms of a particular genotype reproduce successfully then s would be 0.25, i.e. 25% selection against the genotype. An equation can be derived to predict how selection against a recessive allele would change its frequency. This is the simplest situation since only homozygous recessives will show the phenotype upon which selection is operating. The derivation of this and other more complex models can be found in most books on population genetics.

The effects of selection against a recessive phenotype can be predicted from equation 11.12

$$q_1 = \frac{q_o - sq_o^2}{1 - sq_o^2} \qquad\qquad (11.12)$$

where q_1 = recessive allele frequency after one generation of selection
$\quad\quad\; q_o$ = recessive allele frequency in previous generation
$\quad\quad\; s$ = selection coefficient.

This model can be easily converted into a BASIC program. However, a more useful implementation would be to follow the changes over several generations. This would be a simulation of the changes which may be expected to occur in a natural or laboratory population. The data which is produced from this simulation can be compared with published information. In this way the validity and usefulness of the model can be assessed. If the simulation does not fit the observed data then the model and/or the theory may have to be reconsidered.

Example 11.3 uses the facilities of a structured BASIC dialect while Example 11.4 uses the same algorithm but only the universal BASIC facilities are utilised.

Example 11.3

```
10 REM          SELECTION AGAINST A RECESSIVE ALLELE, STRUCTURED DIALECT
11 REM     VARIABLES USED
12 REM          Q - FREQUENCY OF A RECESSIVE ALLELE
13 REM          SELECT - SELECTION COEFFICIENT
14 REM          MAXGEN - MAXIMUM NUMBER OF GENERATIONS FOR THE SIMULATION
15 REM          GENERATION - CURRENT GENERATION NUMBER
20 REM ------------------- INITIALISATION -------------------------
```

```
21 PRINT "WHAT IS THE INITIAL FREQUENCY OF THE RECESSIVE ALLELE ";
22 INPUT Q
23    IF Q > 0 AND Q < 1 THEN GOTO 25
                           ELSE PRINT " Q MUST BE BETWEEN 0 AND 1"
24    GOTO 21
25 PRINT "WHAT IS THE SELECTION COEFFICIENT ";
26 INPUT SELECT
27    IF SELECT >= 0 AND SELECT <= 1 THEN GOTO 29
                           ELSE PRINT" S MUST BE BETWEEN 0 AND 1"
28    GOTO 25
29 PRINT "HOW MANY GENERATIONS SHOULD SIMULATION RUN FOR ";
30 INPUT MAXGEN
31    IF MAXGEN > 100 THEN MAXGEN = 100
32 LET GENERATION = 1
40 REM --------------------- THE SIMULATION ----------------------------
41 PRINT "GENERATION","ALLELE FREQUENCY"
42   WHILE Q <= 1 AND Q >= 0 AND GENERATION <= MAXGEN
43        PRINT GENERATION, Q
44        LET Q = (Q - (SELECT * Q^2)) / (1 - (SELECT * Q^2))
45        LET GENERATION = GENERATION + 1
46        REM GOSUB GRAPH PLOTTING SUBROUTINE IF AVAILABLE
47   WEND
48 END
```

Example 11.4

```
10 REM     SELECTION AGAINST A RECESSIVE ALLELE, UNIVERSAL DIALECT
11 REM VARIABLES USED
12 REM     Q - FREQUENCY OF A RECESSIVE ALLELE
13 REM     S - SELECTION COEFFICIENT
14 REM     M - MAXIMUM NUMBER OF GENERATIONS FOR THE SIMULATION
15 REM     G - CURRENT GENERATION NUMBER
20 REM ------------------ INITIALISATION -----------------------------
21 PRINT "WHAT IS THE INITIAL FREQUENCY OF THE RECESSIVE ALLELE ";
22 INPUT Q
23 IF Q > 0 AND Q < 1 THEN GOTO 26
24 PRINT " Q MUST BE BETWEEN 0 AND 1 "
25 GOTO 21
26 PRINT "WHAT IS THE SELECTION COEFFICIENT ";
27 INPUT S
28 IF S >= 0 AND S <= 1 THEN GOTO 31
29 PRINT " THE SELECTION COEFFICIENT MUST BE BETWEEN 0 AND 1"
30 GOTO 26
31 PRINT "HOW MANY GENERATIONS SHOULD THE SIMULATION RUN FOR ";
32 INPUT M
33 IF M > 100 THEN M = 100
34 LET G = 1
40 REM -------------------- THE SIMULATION ---------------------------
41 PRINT "GENERATION","ALLELE FREQUENCY"
42 REM WHILE Q IS BETWEEN 0 AND 1 AND THE NUMBER OF GENERATION IS < M
43 IF Q < 0 OR Q > 1 OR M > 100 THEN GOTO 48
```

```
44 PRINT G,Q
45 LET Q = (Q - (S * Q^2)) / (1 - (S * Q^2))
46 LET G = G + 1
47 REM  GOSUB GRAPH PLOTTING SUBROUTINE IF AVAILABLE
48 REM WEND
49 GOTO 43
50 END
```

Both of these programs are very simple. The algorithm is based upon recalculating a value for q using the frequency in the previous generation combined with the value of s, the selection coefficient. These calculations are continued as long as q is within biologically meaningful limits $(0-1)$ and the number of generations is less than the specified maximum. The output from this simulation would be easier to interpret if it was presented graphically rather than as numbers. This would be relatively simple to achieve using the criteria discussed in Chapter 9.

Sample run

```
> RUN

WHAT IS THE INITIAL FREQUENCY OF THE RECESSIVE ALLELE ? .5
WHAT IS THE SELECTION COEFFICIENT ?.2
HOW MANY GENERATIONS SHOULD THE SIMULATION RUN FOR ? 10
GENERATION   ALLELE FREQUENCY
  1             .5
  2             .473684
  3             .448956
  4             .425809
  5             .404204
  6             .384078
  7             .365354
  8             .347946
  9             .331766
  10            .316725

> OK
```

Population genetics provides many models which are suitable for use in computer simulations. This simulation has used one of the simplest selection models. More complex models can be found in most genetics textbooks.

11.4 Model 3 – competition between two species of animal

Many biological models, including those covered so far in this chapter, are deterministic. This means that chance plays no part in the process being modelled. However, chance is an important component of many biological processes. Consequently models can be made more realistic if chance events can be included in the model. Models which make use of chance, or probabilities, are called stochastic models. When stochastic models are used to predict the behaviour of a system the resulting output is said to be a Monte Carlo simulation. It is possible to include chance events in computer models because there are routines availabe (the RND function in BASIC) which can produce pseudorandom numbers.

This model is based on equations described by Pielou (1974).

The effects of intra- and interspecific competition upon the growth rate of a pair of competing animal populations can be predicted by Gauss's competition equations. These equations are similar to those of the Lokta - Volterra model of animal competition.

The growth of species A is predicted by:

$$\frac{1}{A} \cdot \frac{dA}{dt} = r_a - s_a A - u_a B \tag{11.13a}$$

The growth of species B is predicted by:

$$\frac{1}{B} \cdot \frac{dB}{dt} = r - s_b B - u_b A \tag{11.13b}$$

where A and B are the respective numbers of individuals of the two species

r_a and r_b are the intrinsic rates of increase for species A and B respectively

s_a and s_b are measures of intraspecific competition, i.e. as each population grows it begins to inhibit itself. The growth of either species in isolation could, therefore, be predicted from $r_x - s_x X$, where X is the species under consideration.

u_a and u_b measure the extent to which the growth of each species is inhibited by the presence of the competing species. They are therefore competition coefficients.

Pielou gives values for these variables which were obtained from a classic series of competition experiments that used two species of the flour beetle, Tribolium. Equations 11.13 (a) and (b) become:

$$\frac{1}{A} \cdot \frac{dA}{dt} = 0.100 - 0.0007A - 0.001B \tag{11.14a}$$

$$\frac{1}{B} \cdot \frac{dB}{dt} = 0.075 - 0.0007B - 0.0007B \tag{11.14b}$$

These two equations form a deterministic model of competition between the two species. This model could be included in a BASIC program which would then be used to simulate competition between the two species. However, it is also possible to introduce chance events into the model. This is possible if the probability of a birth or death within each species is known.

135

These probabilities can be calculated if the numbers of individuals and the birth and death rates are known for each species. Pielou describes the methods for deriving the birth and death rates and also gives values for the two species represented in equations 11.14(a) and (b). The values that will be used in this model are given below.

birth rate of species A (b_a) $= 0.11 - 0.0007A - 0.001B$ (11.15a)
death rate of species A (d_a) $= 0.01$
birth rate of species B (b_b) $= 0.08 - 0.0007B - 0.0007A$ (11.15b)
death rate of species B (d_b) $= 0.005$

The probability of any of these events occurring will be related to the number of individuals present. The proportional probabilities for each event will be equal to the rate multiplied by the number of individuals. Thus the proportionate probability of a birth occurring in species A is equal to the number of A individuals multiplied by b_a (Ab_a).

The relative probabilities of each of these events can then be calculated as shown below:

p_a probability of a species A birth $= Ab_a/(Ab_a + Ad_a + Bb_b + Bd_b)$
q_a probability of a species A death $= Ad_a/(Ab_a + Ad_a + Bb_b + Bd_b)$
p_b probability of a species B birth $= Bb_b/(Ab_a + Ad_a + Bb_b + Bd_b)$
q_b probability of a species B death $= Bd_b/(Ab_a + Ad_a + Bb_b + Bd_b)$

Therefore

$$p_a + q_a + p_b + q_b = 1$$

If the birth and death rates given above are used with initial population sizes of 20 (A) and 40 (B) the values for the probabilities are equal to:

Event	Probability	Cumulative sum
p_a	0.368	0.368
q_a	0.066	0.434
p_b	0.500	0.934
q_b	0.066	1.000

The RND function can be used to generate a random number between 0 and 1. Numbers generated in this way can be used to determine which, of the four possible events, will occur at any point in time. Thus in the example shown above, a random number between 0 and 0.368 would be associated with the birth of an individual of species A. A number between 0.369 and 0.434 would be associated with the death of a member of species A, etc. Any of these events will have an effect on one of the population sizes. Since both population sizes are used in the calculation of the birth rates (equations 11.15 (a) and (b)) the birth rates must be recalculated using the new population sizes. This, in turn, will affect the relative probabilities of the four events. They must also be recalculated. A new random number can now be generated and the populations adjusted accordingly. This process can be repeated for a fixed number of times or until one of the populations reaches a specified size. This technique forms the basis for the algorithm used with Example 11.5.

136

The best way of following the changes in population size is by means of a graph. The numbers which are generated by this program could be easily used as coordinates on a graph.

Since few changes would be required to convert Example 11.5 into a program capable of being used with any dialect of BASIC, only one program is provided on this occasion.

Note that a graph plotting subroutine is not shown. The graph, Fig. 11.1, was produced on an Apple II+ microcomputer using GBASIC (CP/M based graphics BASIC).

Example 11.5

```
10 REM        STOCHASTIC MODEL OF ANIMAL COMPETITION
11 REM        BASED UPON AN EXAMPLE BY PIELOU (1974) PP 221 - 227
12 REM     VARIABLES USED
13 REM        A - NUMBER OF SPECIES A INDIVIDUALS
14 REM        B - NUMBER OF SPECIES B INDIVIDUALS
15 REM        DA, DB - DEATH RATES FOR THE TWO SPECIES
16 REM        BA, BB - BIRTH RATES
17 REM        ABA, BBB - PROPORTIONAL PROBABILITIES OF BIRTHS
18 REM        ADA, BDB - PROPORTIONAL PROBABILITIES OF DEATHS
19 REM        TPROB - TOTAL OF THE PROPORTIONAL PROBABILITIES
20 REM        PBA, PBB - RELATIVE PROBABILITIES OF BIRTH
21 REM        PDA, PDB - RELATIVE PROBABILITIES OF DEATH
22 REM        CP1 TO CP4 - CUMULATIVE PROBABILITIES
23 REM        RN - RANDOM NUMBER
24 REM
30 REM ------------------- INITIALISATION ---------------------------
31 REM CLEAR THE SCREEN
32 PRINT "ENTER THE STARTING VALUES FOR THE SIZES OF POPULATION"
33 PRINT " A ";
34 INPUT A
35 PRINT " B ";
36 INPUT B
37 PRINT "ENTER THE DEATH RATES FOR POPULATION"
38 PRINT " A ";
39 INPUT DA
40 PRINT " B ";
41 INPUT DB
42 REM CLEAR THE SCREEN AGAIN
49 REM ----------------- THE SIMULATION LOOP ------------------------
50 PRINT "NO. OF SPECIES A","SPECIES B"
51 PRINT A, B
52 WHILE A > 0 AND B > 0 AND A < 100 AND B < 100
53    LET BA = 0.11 - 0.0007 * A - 0.001 * B
54        IF BA < 0 THEN BA = 0
55    LET BB = 0.08 - 0.0007 * B - 0.0007 * A
56        IF BB < 0 THEN BB = 0
57    LET ABA = A * BA
58    LET ADA = A * DA
```

137

```
59   LET BBB = B * BB
60   LET BDB = B * DB
61   LET TPROB = ABA + ADA + BBB + BDB
62   LET PBA = ABA / TPROB
63   LET PDA = ADA / TPROB
64   LET PBB = BBB / TPROB
65   LET PDB = BDB / TPROB
66   LET CP1 = PBA
67   LET CP2 = CP1 + PDA
68   LET CP3 = CP2 + PBB
69   LET CP4 = CP3 + PDB
70   LET RN = RND (1)
71       IF RN < = CP1 THEN A = A + 1
72       IF RN > CP1 AND RN < = CP2 THEN A = A - 1
73       IF RN > CP2 AND RN < = CP3 THEN B = B + 1 ELSE B = B - 1
74   REM  GOSUB PLOTTING ROUTINE USING A AND B AS THE BASIS
75   REM  FOR SCREEN COORDINATES
76 WEND
77 END
```

Sample run
This sample run is based upon the output from Example 11.5. A simple
graph plotting routine was used to obtain the output shown below.

```
>RUN


ENTER THE STARTING VALUES FOR THE SIZES OF POPULATION
   A ? 25
   B ? 35
ENTER THE DEATH RATES FOR POPULATION
   A ? .009
   B ? .004
```

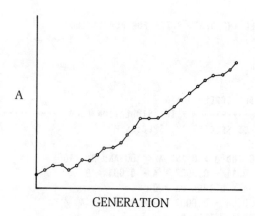

Fig. 11.1 Sample output from the competition experiment model

11.5 Summary

Three simple computer models have been developed by writing programs based on previously published equations (mathematical models). The output from each of these models is reasonably consistent with the expected behaviour of the 'real' system. Although the three models are simple they can still be used as an aid to gaining an understanding of the biological processes upon which they were based. BASIC is not an ideal language for the construction of more complex models. It has two main faults. Long and complex BASIC programs can be very slow to process data. (Methods for improving execution speed are outlined in Chapter 14. However these methods usually reduce program structure.) The second problem with BASIC is its general lack of structure. While this is not a major problem with short programs it can make long programs very tedious to debug and alter. A much better language for the construction of complex computer models is Pascal.

Chapter 12 Information Technology

12.1 Introduction

Information technology, also known by its rather unfortunate acronym, IT, is an amalgam of telecommunications and computer technologies that represents one of the fastest growing areas of science and industry. Information technology will dramatically affect both our private and working life in the very near future. It is difficult to define precisely the boundaries of information technology since it is such a diverse subject. However, there are a number of areas which are going to become important in modern biology and these will be considered in the final part of this chapter. One essential requirement for all information technology systems is that computers must be able to communicate with each other and therefore the first part of this chapter will consider some of the solutions to this important problem.

12.2 Computer communications

If two computers are to communicate with each other two requirements must be fulfilled. Initially, each computer must have an interface compatible with the outside world and the software which will enable them to utilise the interface. At the moment four standard interfaces are commonly used: RS-232, V24, IEEE-488, Centronics. The first two use serial (sequential) information transmission techniques while the last two use parallel (simultaneous) information transmission. The parallel interfaces can only be used to transmit data over relatively short distances because synchronisation between the bits breaks down over long distances (the maximum distance is about 30 m). This may result in a bit from one byte arriving with those of an adjacent byte. In addition, the multiwired cables which must be used for parallel transmissions are not compatible with existing communication channels such as telephone cables. Since the RS-232 and the V24 interfaces use serial transmission they require fewer wires. Consequently, the existing communication channels can be used to transmit information over very long distances. Even if two computers are equipped with interfaces such as the RS-232 and are connected via suitable cables there are still a number of problems that must be overcome before intercomputer communication is successfully established.

Although the ASCII code is, in theory, a standard system, there are some minor but annoying differences between computers and their peripherals

140

particularly with respect to the control characters. Such problems can be corrected by relatively simple programming. It will also be necessary to overcome problems related to the rate and format of the transmitted information. The rate at which each machine is capable of transmitting and receiving information via a serial interface is expressed as a baud rate. The speed of transmission in bauds is equal to the number of signals per second. If each signal is a bit, then the baud rate is equal to the number of bits per second. Unfortunately, this is not always the case and, therefore, baud rates can be difficult to quantify and compare. If we assume that two machines are transmitting and receiving at the same baud rate there are still problems associated with keeping the two computers in step with each other. The two main solutions to this problem are concerned with the way in which the information is transmitted. If a high speed of data transfer is required a synchronous transmission is used. A synchronisation message is transmitted which gets the two computers into step. Synchronisation is maintained by sending regular clock signals with the data. Asynchronous transmission is more common with slower data transfer rates and uses a clock signal which is sent with each byte. This clock signal is in the form of a specific number of prebyte or start bits plus associated postbyte or stop bits.

Additional checks on the accuracy of data transfer can be carried out by using an error checking routine. The simplest of these uses parity bits. All of the common upper- and lowercase alphanumeric and control characters can be represented by the seven-bit ASCII code. This means that there is one bit spare (bit 7) which can be used for data checking; this then becomes known as the parity bit. Two systems, odd and even parity, are used. If even parity is used the seventh or parity bit is only set to one if the byte currently contains an odd number of ones. Consequently, the number of set bits in a byte will always be even. Therefore all of the bytes received should contain an even number of ones. If a byte is received with an odd number of ones the message must have been corrupted during transmission and the receiving computer can request a retransmission of the faulty byte. Odd parity is the opposite, in that bit seven is used to ensure that all transmitted bytes contain an odd number of ones. Data corruption would, therefore, have occurred if a byte containing an even number of ones was received. This system is illustrated in Table 12.1.

Table 12.1 The use of parity bits

Character	ASCII (decimal)	ASCII (binary)	Odd parity byte	Even parity byte
+	43	0101011	10101011	00101011
8	56	0111000	00111000	10111000
H	72	1001000	11001000	01001000
h	104	1101000	01101000	11101000
z	122	1111010	01111010	11111010

Other error checking routines use either checksums or cyclic redundancy checks. Both methods transmit information in blocks. If the first method is used each block is terminated by a checksum which is based upon an addition of the bytes within the block. The receiving computer would carry out an independent checksum calculation. The two checksums are then compared. If they do not agree the block of transmitted information must have been corrupted. The receiving computer can then request a retransmission of the block. The last method is the cyclic redundancy check (CRC). This is the method used by most floppy disk controllers to verify the validity of their recordings. It is a more complex system which has been incorporated into both hardware and software products. The CRC is generated by a complex algorithm which makes use of a polynomial type equation. The CRC takes a value which can satisfy a test equation. The CRC is appended to a block of data and then used in a similar way to the checksum.

If computers are to communicate via telephone cables then each must be connected to the telephone system. The most reliable method is a direct non-acoustic, electrical connection which will allow very rapid data transfer rates while maintaining a high degree of reliability. However, this is also the most expensive method. A less reliable connection, which is also cheaper and slower, uses a normal telephone handset which is connected to the computer via an acoustic modem. A modem is an electronic device which can convert a digital signal from a computer into an audible tone suitable for telephone transmission and vice versa.

12.3 Local area networks

Improvements in the available hardware and software mean that it is now relatively easy to set up links between a number of microcomputers in order that a Local Area Network, or LAN, may be established. The establishment of a LAN within an office block or department can bring about many advantages which include the sharing of expensive peripherals, such as hard disks and letter quality printers. It also allows extensive and possibly expensive information to be shared between many users. Finally, it enables a very fast electronic mail service to be introduced as an alternative to the traditional memoranda. However, as might be expected, there are problems to be overcome before an efficient and reliable network of microcomputers can be established.

There are a number of different LANs in existence which differ in their spatial organisation and operational procedures. They share, however, one common characteristic. The various pieces of hardware, which form the LAN, are joined to the common network at points known as nodes. A node is a physical and logical connection between a device (such as a printer, hard disk or computer) and the network. Each node is identified by an unique address (in the LAN) which is recognised by all of the devices connected to the network.

LANs are organised into one of three spatial arrangements or topologies: bus, ring and star. These are shown diagrammatically in Fig. 12.1. The material connecting the nodes is known as the medium and in most cases consists of a type of electrical cable although there are experimental systems which use fibre optics. In this latter case problems are encountered in joining a device to the network. The rate of information transfer through the medium depends upon the type of transmission used. Broadband transmission, which uses radio-frequency signals, is capable of the highest rates, while the less complex baseband transmission, using a two-state signal, is the slowest. Broadband transmissions require very expensive hardware and such

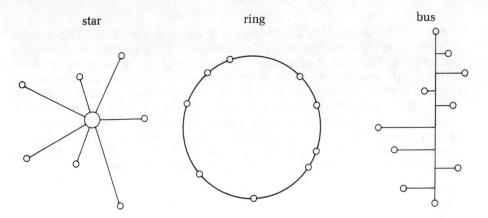

Fig. 12.1 Local area network topographies

systems will probably be restricted to networks which are established by large corporations and governments.

It is apparent that there must be a mechanism by which one device can attain temporary dominance of the network in order that the chaos produced by all the devices 'talking' at the same time is avoided. The mechanism by which this is achieved is known as the access protocol. The simplest access protocol involves the use of a single master device which delegates temporary control to other nodes. This is the system used by the star configuration. A second mechanism which can be used with bus and ring networks involves the use of a token. The device which has possession of the token has control of the network. The third possibility is the CSMA-CD method, an abbreviation for the confusingly named Carrier Sense Multiple Access – Collision Detection. With this method, a device waiting for access to the network sends out a signal and waits to see if it returns uncorrupted. If this occurs the network is free and data may be safely transmitted. If the test signal does not return intact then it must have collided with data from an existing user and thus the network is busy. If the network is busy, a device will wait for a random period of time before sending another signal. There is a related method called CSMA-CA (Collision Avoidance). CSMA-CA systems are less expensive because they do not need the hardware which is used to monitor transmissions. They use complex algorithms and error checking routines to avoid data collision problems. The final method, which is available only on ring networks, uses a system of 'packets'. A reasonable analogy for this system is a ring of railway track around which a number of railway wagons are continuously circulating. If a wagon or packet is empty it can be filled by any device and emptied at its destination node. This means that a particular set of data may be broken up between non-consecutive packets but this does not matter since each packet is headed by the address of its destination node.

There is as yet no standard type of network, although there are currently two major commercial systems. These may be overwhelmed when IBM introduces a standard for their very successful microcomputer. The first of these commercial networks is usually known as an 'Ethernet' type. Ethernet uses a CSMA-CD bus topography which is capable of transmitting information at rates of up to ten million bits per second. It has the advantage of

143

support from major manufacturers such as Intel and DEC. The second category uses a system known as the Cambridge Ring which uses the empty packet technique and is also capable of reaching transmission rates of ten million bits per second. The Ethernet system can transmit one and a quarter million characters per second between two devices. However, while this is happening no other device can gain access to the network. In comparison, the Cambridge ring system is capable of transmitting a quarter of a million characters per second between several pairs of devices simultaneously. Current opinion suggests that the most important LANs in the future will be either the Ethernet or token ring systems.

12.4 Databases

A database is a collection of information which is organised in such a way that various specialised programs are able to make use of it. This short description does not indicate the level of complexity associated with the most useful databases, or the great value and flexibility which they can provide. Databases form the cornerstones of the facilities which are going to become increasingly common and important to all biologists as the effects of the information technology boom become more established.

Essentially, a database system is a computer program and its associated databases which offer a number of facilities concerned with the storing and processing of information. In particular it should be able to act as a link between a block of files. There are many database systems on the market whose facilities and cost cover an enormous range. The best database systems for microcomputers currently cost several hundred pounds to purchase and if you are using one that cost much less then you cannot hope to have the same facilities available to you. Computer database systems are only of value when they are used with large sets of data. There is not much point in keeping the addresses of your friends in a computer database since it would be faster to look up their addresses in a more primitive form of database known as an address book!

The information stored in a database is organised around the concept of a data structure (see Appendix E for references). As described in Chapter 8, a complete set of information can be known as a file. A database will normally contain a number of related files each of which can be broken down into blocks called records. The records are, in turn, broken down into smaller blocks called fields. The use of a database is best illustrated by a simple example. If we were to carry out a survey of an area of vegetation we would probably proceed by taking measurements of the vegetation and of certain physical and chemical characteristics to be found in a number of randomly placed quadrats. Eventually we would have a large table of data which lists the characteristics for each of the quadrats sampled. This would be our file of information. An obvious way of breaking down the information would be to consider each quadrat separately. The quadrats are the blocks or records that together make up the file. If we now examine the information recorded for each quadrat or record we can see that it is made up of a number of discrete units of information such as soil pH, soil type, number of plants of a certain species. Each of these units repesents a different field. The structure of a file can therefore be represented as follows:

FILE NAME: Vegetation Survey

Record no.	Field 1	Field 2	Field 3	Field 4 ...	Field n
Quadrat no.	pH	soiltype	soildepth	slope	total cover
0001	6.3	loam	25	10	100
0002	6.4	loam	27	16	98
n	7.2	sandy	10	23	40

Once information has been entered into the database it should be possible to manipulate it in a number of ways. If a database is to be of value it must be capable of performing at least four functions:

handling a variety of data inputs and edit commands
selection and sorting of the records and fields
processing of the information, including simple arithmetic on numeric data
generating reports based upon the information contained within the database.

In the above example the information was entered in both numeric and alphanumeric forms. It is possible to restrict the data to numeric inputs if a numeric key is developed. Thus in the soiltype field a 1 could mean a loam soil, 2 a sandy soil and so on. The disadvantage of such a system is that the brain must translate the numbers into words every time the information is used.

The two most important database facilities are probably the selection and processing facilities. They allow the information contained within the database to be manipulated and processed. This is possible because the database will have established cross-references between files. It should be possible, for example, to request that the database selects those records in which the soil type was sandy. If the database is powerful enough, it should be possible to base the selection on a number of combined fields so that we could identify those quadrats (records) which had a sandy soil greater than 15 cm deep and a slope of 15 degrees or more. These kinds of operations are possible by hand but the time required is prohibitive. In addition to the search and select functions, the database should also be able to sort the records using the information contained in one or more of the fields. A bibliographic database could, therefore, be sorted into alphabetical order using the record field containing the author's name, or it could be sorted into a chronological sequence if the date of publication field was used. A combination of these two fields would rapidly assemble a jumble of references into the bibliographic sequence used by most scientific journals. A facility which is available on the best database programs is one which allows the user to carry out arithmetic operations on one or more fields. In the vegetation survey example this would allow the user to calculate mean values and possibly more complex statistics on fields containing numeric data such as the soil pH. The final facility which should be offered by database programs is one which allows for the construction of data summaries using a format, within the limitations of the program, desired by the user.

One of the most exciting developments of the last few years has been the development of large commercial databases which are available to the general public. In the United Kingdom there are now two databases available via normal domestic television sets which have been slightly modified. CEEFAX is the BBC version and ORACLE is the IBA system. The facilities offered by these two domestic systems are restricted because communication is one-way, from the transmitter to the television set. Both work by storing the information in a number of pages which can be displayed on the television screen. The information is carried, using spare capacity present on the broadcast signal, in the form of a continuously cycling format. The local receiver, or television set, is 'told' which page to freeze and display. This display will not be changed until the next requested page comes round again on the transmission cycle. This type of limited, transmitted database is known as Teletext or Broadcast Videotext.

These commercial databases become more useful when communication is two-way and the user is able to interact with the database. If two-way communication is to be established links must be set up between the host computer and a local terminal, which could be a normal microcomputer equipped with a suitable interface. Broadcast transmissions are impractical for the local terminal so the communication is usually established via telephone cables. The user dials a telephone number which connects to the host computer. Once the telephone connection has been established the two computers can hopefully communicate with each other via a modem. This type of transmitted database is known as Interactive Videotext or Viewdata. A good example of such a system is the PRESTEL service offered by British Telecom. The user pays a small annual rental fee, plus the cost of telephone calls, and is then allowed access to a number of public and private databases. Some of these are free, but others must be paid for on the basis of time connected. The big advantage of the Viewdata systems is that the user can transmit information to the host computer and could, therefore, order and pay for goods, search a database using selected criteria, or even request transmission of computer programs from a library of such material.

Although such systems are of general value they are not specifically aimed at the scientist. There are, however, a number of databases whose 'audience' consists of people interested in particular disciplines. Biologists are lucky in that they are quite well catered for by existing systems and will presumably benefit as other databases become more generally available. Access to these databases is available through online information retrieval systems which are now available in many large libraries.

12.5 Online information retrieval

Online information retrieval systems make use of the facilities that have been described above. A connection is made between a host computer, which carries a number of databases, and a computer terminal which can be many thousands of miles away. Once the connection has been established the user is able to interrogate the database directly, obtaining the required information very quickly. Such systems are going to become very important because it is becoming impossible for libraries to keep in stock all of the material published on even relatively narrow subjects, given the increase in cost and volume of printed material which is appearing every year. Also, there has been a rapid proliferation in the number of scientific journals, so that even if the journals are readily available, it can be too time consuming for one individual to search through the literature. This is particularly so if a search is required over a number of years. The task of literature searching has been considerably eased by the creation, and availability, of bibliographic databases such as Biosis Previews. This contains over 2,500,000 stored items obtained from *Biological Abstracts* and *Bio Research Index*. Subsets of these large bibliographic databases can now be purchased to provide a regular update of recent publications within an area of interest. These areas of interest can be specified by the user or can be one of the many pre-prepared topics such as Carcinogenesis or Pesticides. This information is provided on a magnetic medium, either tape or disk, which can be read by the user's computer. In addition to bibliographic databases there are a number of other biological databases which have more restricted coverage. The following list demonstrates the range of available databases and the subjects they cover.

Agris	An international agricultural database organised by the FAO.
Cancerline	This is actually three related databases produced by the American National Library of Medicine and covers bibliography, current research programs and treatments for cancer.
Chemdex	A dictionary of chemical names and nomenclature derived from *Chemical Abstracts*.
EPB	Environmental Periodicals Bibliography database which covers most aspects of environmental science.
Medline	A bibliography of medical literature maintained by the US National Library of Medicine.
Toxline	Another database produced by the US National Library of Medicine, which covers a range of topics, including pharmacology, drug chemistry and aspects of pollution concerned with toxicology.
ZR Online	A bibliographic database which is equivalent to the publication *Zoological Record*. It is concerned with publications in which animals (humans are excluded) form the central focus.

Access to these and other databases is available via host systems. Hosts are public bodies or private companies which make the databases available through their computers. Users are either charged for time used with the system or they pay a fixed rental charge for use of the facility. Most of the major databases are available through a number of different hosts. The host systems that a biologist is likely to come into contact with are:

BLAISE	British Library Automated Information SErvice. In addition to acting as a host for databases such as Medline, it is also a database in its own right, since it contains details of all books published in the UK since 1950 and all books published in the USA since 1968.
DIALOG	This is the host system of the world's largest collection of databases which are provided by the Lockheed Corporation in the USA.

DIALTECH This is a host system provided by the European Space Agency and allows access to twenty, mainly scientific, databases from its headquarters in Italy.

The use of one of these systems is best illustrated by an example.

An initial telephone link is established with the host computer via a telephone modem. The two computers may be many thousands of miles apart, in which case a satellite link will probably be used. The host system will only allow access after a valid password has been entered. This ensures that the user can be invoiced for the service! If the host has several databases available one of these will be selected by the user who now interrogates this database for the information required. This interrogation is achieved by the use of keywords. The database will search through its records for occurrences of the keywords. Even though the database may contain hundreds of thousands of records the search will usually take only a few seconds. Delays are more likely if there are many users currently attached. The systems usually operate by spending a fixed time period with each user before passing on to the next user. This will not be apparent to the user unless the system is connected to a large number of users.

Suppose that we are currently involved in a research project which is trying to identify which parts of a substrate molecule are important in promoting efficient enzyme–substrate interactions. We are particularly interested in the enzyme hexokinase, since it shows a range of substrate activities. How much work has been carried out on this topic? If we attach to a suitable bibliographic database we can begin to find the answer.

The first step is to identify the keywords. Obviously hexokinase will be one; another possibility is structure, since this could relate to both enzyme and substrate. Since we are interested in the interaction between the two the third keyword could be enzyme–substrate. It is quite possible that X-ray crystallography has been used to investigate the structure of hexokinase so we can use X-ray crystallography as our final keyword.

We now request the database to search its records for the occurrence of our first keyword — hexokinase. Since there has been a lot of work which uses this enzyme we will probably be told that it occurs in several hundred records. Next we can ask how frequently our second keyword, structure, occurs. This is a very general term and it is likely to be present in very many records. However, we can now narrow the field by asking how many times the two occur together. This will immediately reduce the number of records under consideration and we may be down to 10 or 20 records. They probably contain some unwanted material, for instance, there may have been some work on the structure of cells and how hexokinase was localised within the organelles. We could narrow the search down even further by asking how many times the four keywords occur together. Unless someone has used X-ray crystallography to study enzyme–substrate interactions in the enzyme hexokinase, the answer will be zero records. Fortunately the databases usually allow the use of other logical operators (in addition to AND) such as OR and NOT. The keywords can, therefore, be used in logical combinations to search the database. In the hexokinase example a search on the keywords hexokinase and enzyme–substrate and (structure or X-ray crystallography) would identify those records which contained both of the first two keywords and either of the last pair. Note that the use of brackets forces the two words to be considered together. The interrogation operation is shown below.

Enter the keywords

     1. hexokinase
367 records
     2. structure
2781 records
     3. 1 and 2
15 records
     4. enzyme–substrate
132 records
     5. X-ray crystallography
46 records
     6. 3 and 4 and 5
0 records
     7. 1 and 4 and (2 or 6)
2 records

Note that once a keyword or combination of keywords has been used it can be referenced by number. When the relevant records have been identified most databases have a facility which will allow an abstract of the records to be displayed. If a large number of relevant records have been identified it is usually less costly and more convenient to have a printout of the records. Most of the commercial databases have this facility available. The printout will be prepared after disconnection from the database and it should arrive by post several days later.

12.6 DNA databases

Recent advances in restriction enzyme technology and DNA sequencing techniques have allowed rapid advances to be made in the area of molecular genetics. This has produced a large volume of information relating to DNA base sequences. The most efficient way of storing and using this information is to place it in a database which can be accessed by a large number of researchers. There are several databases which are related to this area. The largest example is the NIH GenBANK database. (This is a comprehensive National Institutes of Health database which contains information about the base sequence, including promoter and gene regions, and associated literature. This database is frequently updated to take account of the many advances which are being made in this area.) Other examples include the EMBL (European Molecular Biology Laboratories) database of DNA sequences and the NBRF (National Biomedical Research Foundation) protein sequences database. The background to these new DNA sequencing techniques, and the value of the molecular databases, is briefly described below.

Class II restriction enzymes are a group of enzymes which cut double-stranded DNA molecules at specific sites. An example is the enzyme Hind III. This enzyme cleaves a DNA molecule between the two adenine bases in an AAGCTT sequence. Note that this is an inverted repeat sequence, i.e. its complement on the other DNA strand is TTGCAA. Hind III does not, therefore, produce a clean cut. It leaves behind a 'sticky' end of a single-stranded sequence AGCTT. If a DNA molecule is subjected to the action of a restriction enzyme it will be broken into a number of fragments. The number and size of these fragments will depend upon the frequency and distribution of cleavage sequences within the molecule. These restriction fragments can be separated by Gel electrophoresis (the distance moved by a restriction fragment during electrophoresis is a function of its molecular weight). This technique allows relatively crude comparisons to be made between DNA sequences. If two DNA molecules are identical they will produce identical restriction fragments irrespective of the restriction enzyme used. It is possible to build up restriction maps of DNA molecules. These maps are based upon the points of cleavage made by a number of restriction enzymes.

A more detailed investigation of the DNA can be made by isolating the DNA restriction fragments from the gel. The isolated fragment can be introduced into a vector (a plasmid or viral chromosome). This allows the fragment to be cloned (multiple copies are produced). After the fragment has been cloned it can be sequenced using one of the modern methods such as the Maxam and Gilbert technique.

Once the sequence has been produced it can be searched for the occurrence of particular sequences such as the Goldberg–Hogness or TATA sequence. This would appear to be a promoter sequence for eukaryotic genes which is normally found about 30 base pairs upstream of the CAP sequence (this is an important sequence which ensures that the mRNA binds correctly with the ribosomes). Comparisons can also be made between the base sequences of related genes. A recent application of this technique can be found in a paper by Schmale and Richter (*Nature* (1984) Volume 308, Number 5961, pp 705 – 709). They were able to demonstrate that the gene for the vasopressin precursor in rats with hereditary hypothalamic diabetes insipidus was deficient in one G residue in the protein-coding sequence of the gene. A consequence of this is that a reading frame mutation occurs to produce an incorrect amino acid sequence. The relevant parts of the normal and diabetes DNA molecules are shown below.

Normal	G G A	A G C	G G A	G G C	C G C
Diabetes	G G A	A G C	G A G	G C C	G C T

Note how the missing G residue produces the frame shift mutation. The two sequences of over 1000 base pairs are identical apart from this single deletion.

Obviously these comparisons, and the identification of important sequences, could be made by eye. However, there is a rapidly growing number of base sequences available for comparison and analysis. Libraries of DNA base sequences can be established in computer databases and then subjected to analysis by programs similar to those used with amino acid sequences in Chapter 7.

IntelliGenetics have recently released a comprehensive package of a 32-bit computer plus the software required for the analysis of DNA sequences and restriction maps. The system has two software packages, QUEST and CLONER, plus a database, all of which are controlled by a UNIX operating system. QUEST is the program which is used to search DNA base sequences for various combinations of base patterns. This program allows more powerful search operations to be undertaken than would be possible using BASIC. CLONER is described as a CAD (computer aided design) tool which can be used to design and create new molecules by DNA recombination techniques!

12.7 Summary

Biology is going to be affected by the advances in computers and telecommunications which form the basis for the information technology 'revolution'. Even biologists who have no desire to learn computing are going to find it difficult to avoid the ramifications of IT. There are many other aspects of information technology which have not been covered in this chapter. In particular there are the Expert Systems. These are 'intelligent' databases which use artificial intelligence techniques to produce judgements on facts which have been made available to the program. Two areas in which expert systems are well advanced are medical diagnosis and geological prospecting. There are many other areas of biology in which such systems could be of great value. An expert system which was capable of identifying organisms would find many willing users!

Chapter 13 Working with Mainframe Computers

13.1 Introduction

If you are familiar only with microcomputers you will probably encounter some difficulties when you first use a mainframe computer. Normally a mainframe computer will allow access only to registered users. There are many reasons for this which are concerned primarily with economics and data security. Mainframe computers are very expensive machines to purchase and maintain. Therefore, users must 'pay' for any processing time which they use. All registered users will have a unique user identity code. This must be entered, and accepted by the computer, before any computing can be undertaken. The computer will have an accounts program which monitors the CPU time used by each user. Users can then be invoiced as required. In educational establishments the accounting is normally notional in that most users will not be required to pay for their computing. A mainframe computer will probably have hundreds of registered users, all of whom will have files containing programs and data. It is very important that unauthorised users do not gain access to these files. Imagine how you would feel if a datafile containing your research data was deleted or corrupted, maliciously or accidentally, by another user. Some datafiles may contain confidential information whose contents should only be accessible by a small number of users. Because of these problems most mainframe computers are provided with elaborate security systems (as part of the operating system). There will usually be a password system and a hierarchy of file access control commands. Thus, a user will have an identity code and a password. Individual files can be assigned access controls. These will restrict the rights which users have for files other than their own. For example, a user may be allowed to run a program file but will be prevented from amending or deleting the program.

Mainframe computers also differ from microcomputers in that they are designed to service, simultaneously, many users who may be occupied with completely different activities. Some will be editing files while others are executing programs in a range of programming languages. Usually this multitasking, multiuser element of mainframe computers is transparent to the user. This is because mainframe computers have very complex operating systems which facilitate these operations.

Most mainframe computers will have a wide range of programming languages and program packages available for users. For example, at Manchester Polytechnic the main computer system is based upon a Prime 750 computer linked to a Prime 850 computer. Ten different high level languages

152

are available including BASIC, Pascal, FORTRAN, ALGOL, PILOT and COBOL. In addition there are a large number of program packages associated with such diverse activites as file manipulation and computer models of the national economy.

Mainframe computers are able to support such a diverse range of activities because they are very much more powerful than microcomputers. Most of this extra power results from the more powerful CPUs which are used in mainframe computers. Mainframe CPUs have more powerful instruction sets and execute many more instructions per second than the microcomputer CPUs. James (1982) quotes a number of benchmark timings for a variety of computers. These are obtained by running a set of benchmark, or test, programs. The benchmark timings can be used as a crude measure of computer performance. Although the information is a little dated the conclusions are still valid. Microcomputers, using interpreted BASIC, took over four hours to execute the programs. If machine language was used this decreased to about ten minutes. An IBM 370/148 mainframe computer executed the same programs in less than a minute!

A common problem with mainframe computers concerns the documentation. It appears to be almost universal that such documentation is written in the most incomprehensible dialect of 'computerese'. This is because it is usually written by the systems programmers who manage the computer system and who consequently have an intimate knowledge of its workings. They appear not to comprehend the level of ignorance which is possessed by most of their users!

Interactive computing is undertaken at a terminal, which is similar to the normal microcomputer system but lacks the processing hardware. The terminal consists of an input device, the keyboard, and an output device, the VDU. All computing operations are carried out by the computer which may be several miles away. The public telephone system may be used to link the computer with its terminals. Most mainframe computers also allow batch processing. In this case the job, which may be data entry or a program run, is placed on a set of punched cards which are submitted to the computing centre. They will be entered, via a card reading machine, by one of the computer operators. All output from the program may be directed to a fast line printer. After a length of time, which depends upon how busy the system is, the output can be collected from the computing centre.

Because of the factors discussed above, starting off on a mainframe computer can be much more difficult than switching on a microcomputer. An example of the steps required to run a BASIC program on the computer system at Manchester Polytechnic is given below. Although other computer installations will have different procedures they will utilise a similar set of operations.

Step 1   Find a free terminal and switch on the power.
Step 2   Select which computer is required (Prime 750 or 850).
Step 3   Login (gain connection by entering a valid user identity code and password).
Step 4   Select the BASIC language by entering BASICV.
Step 5   Call up the program with the command 'OLD program name'.
Step 6   The program may now be executed with the normal command 'RUN'.

Leaving the computer is also more complex. You can't just switch off the terminal. This would leave you still connected to the computer and would create problems for the next person to use the terminal. To exit from the computer it is necessary to leave BASIC ('QUIT') and then LOGOUT. This termination command is important because it frees any space (in memory or on disk) currently allocated to you. It also allows the computer accounts program to update its records.

One of the greatest benefits to be derived from mainframe computers is that they usually provide a number of very powerful program packages which can be of great value to the biologist. In particular there are several statistics packages which have been frequently used in biological research. Two of these packages, MINITAB and SPSS, will be used to illustrate the facilities which might be expected from a mainframe statistics package.

13.2 MINITAB

MINITAB is a collection of statistics programs which can be used interactively or in batch processing mode. It is a very simple package to use and requires very little knowledge of computing. Most of the commands are entered as standard English language statements. MINITAB is written in FORTRAN IV, although the user need not be aware of this, and it can be easily implemented on a wide range of medium to large computers (including some 16-bit microcomputers). The developers of this system have made the system available to anyone, with suitable computing equipment, for a nominal fee which covers any costs involved. Further information about the MINITAB project can be obtained from:

Professor T.A. Ryan, Jr.,
Minitab, Inc.,
215 Pond Laboratory,
University Park, Pa. 16802.

There is also a student guide to MINITAB. Details of this can be found in Appendix E.

MINITAB is organised around a worksheet. The basic unit is a column of numbers. The size of the worksheet, i.e. the number of rows and columns, depends upon the space available. Typically it consists of 50 columns each of which has a default maximum length of 200 rows. There will also be a restriction on the number of locations which can be used. Thus, although a 50 × 200 worksheet has 10000 locations there may be a system limit of 5000 locations. This would mean that only 25 of the columns could be completely filled. It would, however, be possible to put 100 numbers in all 50 columns.

MINITAB has a library of words which it recognises. These words are all four characters long. The main requirement for all MINITAB instructions is that the first four letters on a line of instructions form one of the library words. If this is not the case an error message will be printed. If a line begins with a legal word all of the following characters will be ignored until another command character is encountered. This allows commands to be entered as meaningful statements. Example 13.1 demonstrates one of the methods for entering data into a column.

Example 13.1 Data entry
(All words in uppercase have been typed in at a terminal; lowercase indicates a message from MINITAB.) In this implementation MINITAB prompts the user with a short message followed by the > symbol.

```
mtb > READ THE FOLLOWING DATA INTO C1
data> 23
data> 24
data> 103
data> 9
data> 46
data> END
      5 rows entered
mtb >
```

READ is one of MINITAB's library words. It instructs the program to accept a sequence of numbers and place them into one of the columns, which is specified by its number (column 1 in Example 13.1). The end of the data is identified by entering another library word, END. The command could be shortened to READ C1 since these are the only characters recognised by MINITAB in this command line. Once data has been placed in a column it can be manipulated in a variety of ways. If simple statistics, such as the mean and standard deviation, are required they can be obtained by entering DESCRIBE Cn (where n is the column number containing the data). Again, since only the first four characters are used by MINITAB this command could be shortened to DESC Cn. This command is illustrated in Example 13.2.

Example 13.2 Descriptive statistics

```
mtb > READ C1 C2
data> 10 21
data> 12 26
data> 14 27
data> 16 30
data> 18 32
data> 20 36
data> END
      6 rows read
mtb > DESCRIBE C1 AND C2
                c1         c2
n                6          6
mean         15.00      28.67
median       15.00      28.50
tmean        15.00      28.67
stdev         3.74       5.20
semean        1.53       2.12
max          20.00      36.00
min          10.00      21.00
mtb >
```

A large number of functions are available. Thus, the natural logarithms of numbers in a column can be obtained with the command LOGE. Six trigonometric functions are available and operate in a similar way. The sines of the numbers in a column can be determined with the command SIN.

A wide range of statistical tests is available and can be carried out with remarkable ease. Example 13.3 demonstrates how a regression analysis would be performed. Regression analysis is used to obtain the line of best fit through a scatter of points. The regression program provided by MINITAB is very powerful since it allows multiple regression analysis to be carried out. This is used to fit planes to three or more dimensional graphs.

Example 13.3 Regression analysis
(using the data entered into columns 1 and 2 in the previous example)

```
mtb > REGRESS C2 ON 1 SET OF X VALUES IN C1

the regression equation is
c2 = 8.10 + 1.37 c1

                                 st. dev.      t-ratio =
column      coefficient          of coef.      coef/s.d.
            8.095                1.763         4.59
c1          1.3714               0.1146        11.97

s = 0.9587

r-squared = 97.3 percent
r-squared = 96.6 percent, adjusted for d.f.

analysis of variance

due to          df        ss        ms=ss/df
regression      1         131.66    131.66
residual        4         3.68      0.92
total           5         135.33

mtb >
```

Sophisticated and powerful program packages such as MINITAB could not be used on most microcomputers, the majority of which are too slow and do not have sufficient storage available. It would also be very difficult to write such a package in BASIC. Although possible, this would be so slow as to be almost useless.

13.3 SPSS–X

SPSS–X, or SPSS, is a more powerful package than MINITAB, but it is also harder to use. (SPSS–X is a trademark of SPSS Inc. of Chicago, Illinois, for its proprietary computer software.) SPSS is an abbreviation for the rather misleading title Statistical Package for the Social Sciences. Despite its title all of the programs contained within the package can be used in a biological context. SPSS is also written in FORTRAN, but this package demands a greater knowledge of computing than does MINITAB. It is particularly important to understand something about the nature of data formatting in FORTRAN. SPSS is useful because it provides, in addition to the usual basic statistical tests, a number of advanced statistical procedures which can only be undertaken with the aid of a computer. These procedures include multivariate analysis of variance, discriminant analysis, factor analysis and time series analysis. An introduction to SPSS is provided by Norusis (see Appendix E). If you intend to use SPSS you will also need to understand how it is implemented on your computer system. The following sections will introduce FORTRAN formatting and provide an example to demonstrate how SPSS is used.

13.3.1 FORTRAN formatting

In FORTRAN data can be described by its type (alphanumeric or numeric), number of characters and, if relevant, the position of the decimal point. Each item of data is said to have a field width which is equal to the number of characters in that data item. Thus 127 has a field width of 3 while 3.1416 has a field width of 6 characters (including the decimal point). FORTRAN makes an important distinction between real numbers and integers. Real numbers are stored in a floating point format (see Appendix A). Therefore, the description of a number should also indicate the numeric type. F is used for real numbers and I for integers. All of these descriptors are combined into a format descriptor. Some examples are shown in Table 13.1.

Table 13.1 Examples of format descriptors used in FORTRAN

Number	Format descriptor	Notes
127	I3	An integer of 3 characters
8972	I4	An integer of 4 characters
3.1416	F6.4	A real number of 6 characters, 4 decimal places
127	F3.0	A real number of 3 characters, no decimal places
3254.78	F7.2	A real number of 7 characters, 2 decimal places

Often data is entered in batches. For example, 30 weight measurements may be entered. If these weights have different character lengths, for example 98.60 and 122.76, it is not necessary to have a separate format descriptor for each number. A format descriptor is used which will accommodate the longest character string which will be encountered for a particular data item. In the weight example both data items could be described by F6.2. The 98.60 data item would be right justified by inserting a blank space before the 9. This blank space would not alter its numeric value but the string of characters would now be 6 characters long.

When FORTRAN was developed most jobs were run in a batch processing mode with the data and programs being entered on cards. The card specification became standardised at a width of 80 columns. Many of the features of FORTRAN are designed around the format of this card. Therefore, even when a FORTRAN job is being controlled from a terminal, in an interactive mode, the program and data are still structured around the 80 column format. It was often easier to put many separate data items onto one card. This data is split into a number of fields of variable length. The organisation of the data on these cards is described by a FORMAT statement. If several fields have the same structure, for example F3.1, this information can be presented in abbreviated format. If data on a card consisted of 3 fields of data with an F3.1 format plus two fields containing integers 4 characters long, the card format would be 3F3.1,2I4. Spaces, i.e. columns which do not contain data, are indicated by X, or nX if there is more than one space (n = number of spaces).

13.3.2 The SPSS job

An SPSS job normally consists of three files. The command file contains a description of the data, with information such as its location and format. It also contains details of the analysis to be carried out. Although data can be included in the command file it is more usual to place it in a separate datafile. This datafile can have any structure which can be described in FORTRAN formatting terms. The third file is the output file which allocates a name to the file which will receive the output. Thus, the output will be empty until after the program has been run and its contents can only be inspected by listing the output file.

The command file consists of a list of program lines, each having a maximum length of 80 characters. Overflows can be continued onto the next line. The program lines are split into two fields. The first, the command field, occupies columns 1–15. The second field is the specification field and occupies columns 16–80. Only valid SPSS commands can be placed in the first 15 columns; anything else in that field will cause an error. Errors are reasonably common with inexperienced users but fortunately SPSS supplies a relatively comprehensive description of any errors which have occurred. The next section is an example of a complete SPSS job.

An added complication with SPSS is that the command and datafiles must be created using a text editor if the job is being executed interactively. Therefore, before you can run an SPSS job on a mainframe computer you will need to know how to use the system editor.

13.3.3 SPSS example

The following example was prepared and executed on a Prime 750 computer at Manchester Polytechnic. This example demonstrates how it is possible to use a relatively complex statistical procedure, discriminant analysis, with the SPSS system. One of the uses of discriminant analysis is to identify which variables are responsible for most of the differences between two groups. This technique has many existing and potential biological applications. The example discussed here concerns identifying those parts of the substrate molecule which are important in determining hexokinase activity. Hexokinase is an enzyme which can utilise a range of D-sugar substrates. However, it is not active with all such compounds, even though they may have similar molecular structures. If two groups, substrates and non-substrates, are compared it may be possible to identify those molecular features of the substrate which are important in determining hexokinase activity. Thus the two groups will be discriminated on the basis of a number of molecular characteristics. Note that before discriminant analysis can be carried out it must be possible to identify and describe individuals from the groups under investigation. (Another application of discriminant analysis is allocating unassigned individuals to their most probable group.) In this example the substrates and non-substrates were described by the presence or absence of molecular features, for example the positions of any -OH groups. Thus, a profile of binary characteristics was obtained for each compound. Compounds were assigned to the two groups on the basis of published information about their activities with the enzyme. It is important to realise that this activity information is not used in the analysis.

Table 13.2 The datafile (called SUBSTRATESTRUCTURE)

01	10011111000001
02	10010111000001
03	10011110000002
04	10001111000002
05	10010111100001
06	10011101010002
07	10011011001002
08	01010111100001
09	10010111000102
10	10010111000112

The format of the datafile is I2,3X,14I1 and consists of a substrate number (I2), three spaces (3X) and 14 integers each of which is one column long (14I1). The first thirteen of these are binary numbers which indicate the presence or absence of a particular molecular feature. The final column identifies which group that substrate belongs to: 1 is an ineffective substrate and 2 an effective substrate. Thus the datafile contains information about ten compounds, four of which are ineffective substrates.

Table 13.3 The command file

```
RUN NAME          DISCRIMINANT ANALYSIS OF HEXOKINASE SUBSTRATES
VARIABLE LIST     A,B,C,D,E,F,G,H,I,J,K,L,M,ACT
INPUT FORMAT      FIXED(5X,14F1.0)
INPUT MEDIUM      [SUBSTRATESTRUCTURE]
N OF CASES        10
DISCRIMINANT      GROUPS = ACT(1,2)/VARIABLES A TO M
OPTIONS           5,6,7,8,17,18,19
READ INPUT DATA
FINISH
```

The first line assigns a name to this job. The second line lists the names of all variables to be used in the analysis. The molecular features have been given dummy names in the form of letters A to M. ACT is the final column containing the activity group. Note that the first variable in the datafile, the substrate number, is not used in the analysis. The next card describes the

159

format of the data in the datafile. The 5X indicates that any data in the first five columns should be ignored, hence the substrate number is not used in the analysis. SPSS does not allow the use of the I identifier and therefore the data is described by 14F1.0 (14 fields of data each of which is one character long with no decimal point). The FIXED descriptor indicates to the program that data is organised in the same format on each line of the datafile. The fourth line identifies the source of the data for analysis. In this case it is in a file called SUBSTRATESTRUCTURE. The next line sets the number of cases, in the datafile, which are to be analysed. The sixth line specifies which analysis is to be undertaken. The structure of this line indicates that a discriminant analysis should be undertaken on two groups labelled ACT1 and ACT2. The variables A to M should be used in the analysis. The next card selects various options which are to be used in the analysis. These control the amount of information which should be sent to the output file. The penultimate card requests the program to start reading the data while the final card indicates that the end of the job has been reached.

When both of these files have been completed the job can be run by entering:

SPSS Commandfilename Outputfilename

Hopefully you will soon get a message that nine cards have been processed and no errors detected. If this has happened the output can be examined by spooling (requesting a printout) the output file. The output file is too long to list here (over 700 lines). It would also be difficult to understand for people without a knowledge of discriminant analysis. The analysis did allow the identification of certain molecular features which are important in determining enzyme–substrate interactions between hexose sugar substrates and hexokinase. In particular, the analysis suggested that hydroxyl groups on the substrate molecule, which have the same orientation as in D-glucose, at carbons 1, 3, 4, and 6 are important in lowering the Michaelis constant. Also, the substitution of an acetyl group at the second carbon significantly decreased the affinity between the sugar and hexokinase. The value of this type of technique is that the analysis and interpretation can be achieved very rapidly.

Chapter 14 Program Optimisation

14.1 Introduction

Programs written in BASIC tend to execute relatively slowly. This is because BASIC is normally an interpreted language. If a BASIC program is performing a simple computation, speed will not be a major problem. However, if the program manipulates large datafiles or if it is interacting with the user in such a way that rapid computer responses are required, the slow speed of execution can be very detrimental.

Fortunately the speed at which programs execute can be increased in a variety of ways. Each of the following options should be considered if speed is important.

14.1.1 Change the programming language

The fastest programs are those written in machine language. It can, however, be quite difficult to write even relatively simple programs in machine language. The process can be simplified if an editor/assembler program is available. This is a software 'tool' which allows machine language instructions to be written as mnemonics.

Many high level programming languages produce programs which execute much faster than their BASIC equivalents. However, these languages, such as Pascal and FORTRAN, are not freely available on microcomputers (where speed will be a greater problem than on mini- and mainframe computers). There is also the not inconsiderable problem of learning a new language.

14.1.2 Use a BASIC compiler

A BASIC compiler is a program which converts a BASIC program into machine instructions. This compiled version is then saved and can be executed when required. Since the program is already in machine language instructions, a compiled program will execute faster than an identical program which executes in the normal interpreted manner. Advertisements for microcomputer BASIC compilers often promise very large improvements in execution speed but a five to tenfold improvement would be about average. Compiling is a time consuming process. Initially, the program will be written using a text editor or word processor. This produces the source code which can be saved as a text file. Any alterations to the source code can only be carried out with the text editor. Once the source code has been completed it can be compiled and linked. Linking is carried out by another program which converts the compiled program into one which can be executed. This involves machine language changes which are beyond the scope of this book. If you are lucky your program will have compiled successfully. More usually the first compilations will produce a list of errors which must be corrected in the source code via the text editor. The program can then be recompiled and linked. An additional problem with some compilers is that they place restraints on the program structure. Many of the less expensive compilers will allow only integer variables and arithmetic to be used in programs.

14.1.3 Optimise an interpreted BASIC program for speed

This is the option which is most freely available. Several methods of optimising programs for faster execution will be introduced in the next section. It is important to realise, even at this early stage, that many of the optimisation techniques destroy the program structure and make programs very difficult to read. Many of these techniques exploit the ways in which the BASIC interpreters execute programs.

14.2 Optimisation of BASIC programs

14.2.1 Improve the algorithm

It is often possible to change or improve the algorithm to produce a faster program. This is particularly true for sort routines. The bubble sort algorithm used in Chapter 10 was selected because it is one of the least complex sort algorithms. It is also very slow. If a bubble sort is replaced by a shell sort algorithm, a long list can be sorted in about a fifth of the time taken by the bubble sort. The more efficient algorithms can often be found in programming books and some of the better computing magazines.

14.2.2 Program design

When a program instruction, such as GOSUB or GOTO, diverts program execution the interpreter does not immediately jump to the new line number. Rather, it returns to the start of the program and searches sequentially through until it finds the line number which has been called. Consequently, if an instruction is to be frequently called it should be placed near to the beginning of a program. The highest numbered lines should be used for instructions which are infrequently used or instructions which control non-speed-critical processes.

14.2.3 Loops

Whenever possible a GOTO instruction should not be placed within a loop. This is because of the problem explained in the previous section. Many programs use GOTO instructions in FOR ... NEXT loops to jump to the NEXT command and thus avoid a block of instructions. These can often be replaced by REPEAT ... UNTIL or WHILE ... WEND constructions, if they are available. This change in program structure has two advantages. The structured commands improve the program structure and decrease execution time.

14.2.4 Initialise variables

When a variable is used in a BASIC expression the interpreter consults a table of variables to determine the variable's position in memory. Variables are added to the table in the order in which they are encountered by the interpreter. The interpreter will locate variables which are near to the start of the list faster than those variables at the end of the list. It can, therefore, be advantageous to initialise variables at the start of the program in a sequence which reflects their importance to execution speed. If a variable is referenced many times during a speed-critical part of the program it should be one of the first variables encountered by the interpreter.

14.2.5 Use integers

If the interpreter allows variables to be defined as integers this facility should be used as often as possible because the interpreter can usually process integers three times faster than real numbers. Integers are stored in a format which is more easily accessed by machine code than is the format of real numbers (see Appendix A for details). It may be necessary to define which variables are integers at the start of the program. Other interpreters recognise integer variables by terminating the variable name with a special character, often the % sign. The execution time of a loop can be decreased if the loop index variable is defined as an integer. Similar improvements can be made for all variables which never hold real numbers.

14.2.6 Avoid numerical constants

Numerical constants are frequently used in programs, for example, 3600 and 981 in program Example 1, Chapter 3. However, if speed is important the numbers should be assigned to variables. If π is to be used many times in a program it will be faster to initialise a variable to 3.1416 at the start of the program and then use this rather than the numeric constant. This is because every time the interpreter comes into contact with a numeric constant it must convert the number into a format which can be used at the machine language level. Initialising a variable to this value at the start of the program ensures that the conversion will occur once only.

14.2.7 Reduce the number of program lines

Processing of each program line carries an execution overhead. Therefore, if the program can be packed into fewer lines it will be executed faster. One of the simplest ways of achieving this is to use multistatement lines. Many interpreters allow more than one statement to occupy the same program line. The BASIC statements are usually separated by colons. The limit is placed by the maximum number of characters which can be used with one line number. This is frequently 255 characters. An example of a single statement per line program and its multistatement equivalent are shown below.

```
10 FOR I = 1 TO N
20 READ X(I)          20 FOR I = 1 TO N : READ X(I) : NEXT I
30 NEXT I
```

Although the multistatement version would execute faster, and use less memory, it is much harder to read and amend the program.

14.2.8 Reduce the number of characters

When the interpreter is executing a program line it first scans each character. Therefore, if the number of characters is reduced the program will execute faster. There are a number of ways of achieving this reduction. All REM statements and spaces can be removed (note that some interpreters demand that spaces are inserted between commands and variables). Also shorter variable names can be used. All of these procedures will vastly reduce program legibility. It is often advisable to keep two copies of a program. One should be readable, but slow to execute, while the other is fast and possibly unreadable.

14.2.9 Select the fastest BASIC commands

It is sometimes possible to achieve the same result with different BASIC commands. If several possibilities are available they are unlikely to execute at identical rates. The best example of this is squaring a number. Consider the following two programs.

```
            A                          B
10 FOR I = 1 TO 1000         10 FOR I = 1 TO 1000
20 LET D = I^2               20 LET D = I * I
30 NEXT I                    30 NEXT I
```

Both programs square all of the integer numbers between 1 and 1000. D is used as a temporary variable to store the results of the computations. In Applesoft BASIC, program A executes in 50 seconds while program B completes the same task in 14 seconds. If alternative methods are available the fastest can usually be identified by writing simple test programs such as those shown above.

Chapter 15 Summary

A biologist is often expected to be a 'jack of all trades'. This is because biologists work with very complex systems in which there will be many interacting variables. Although the chemistry, physics and mathematics of biological systems can be difficult to investigate and understand, a biologist is expected to integrate ideas and concepts from these disciplines into explanations of biological systems. Computers represent a powerful tool which can be used to strengthen and intensify this integration. However, as stated by Barrett (*The Times Higher Educational Supplement*, 17.6.1983) computing has developed relatively slowly in the biological sciences compared with other scientific disciplines. There is evidence that this is now changing.

An indication of the potential importance of computers in biology is demonstrated by the fact that the Department of Education and Science recently organised a national conference to discuss the role of microcomputers in biology in higher education (Oxford Polytechnic, 9–13 April 1984). This was the first national conference called to discuss the impact of microcomputers in any of the higher education scientific disciplines. Three main areas of current usage were identified. Surprisingly the most common application was word processing. A user can become proficient with a word processing system and yet have very little understanding of programming and computers in general. This could be important since it may provide an introduction, and the incentive, for biologists to become involved with computing. The other two main areas of microcomputer usage in biology were research and teaching. In general these two areas had similar requirements which were basically 'number crunching' and computer graphics. A major problem appears to be a shortage of good biological software for microcomputers. There are many reasons for this but the main one would seem to be that there are few people who are good programmers and who possess a sound biological background. One of the recommendations of this conference was that some of the new information technology, in the form of a Prestel database (see Chapter 12), should be used to coordinate software developments throughout the country. It would appear that there is an important role for computers in biological education. All that is required is more money and better programmers!

Computers of all sizes are already making an important contribution to a large proportion of current biological research. One impact has been that complex procedures, such as advanced statistical tests, are now readily available and are being used. There is, however, a danger which must be recognised. There is a computing proverb called GIGO, an acronym for garbage in – garbage out. It is very easy to enter numbers into a program and get an answer. It should be remembered that the quality of this answer is only as good as the data which produced it. The increased availability of complex analysis procedures brings with it a need for an increased understanding, by the biologist, of the underlying principles of the analysis. The converse is true for the computer scientist who may be writing computer simulations and models of biological processes. The quality of these programs will not only depend upon how well the program is structured but also upon the underlying biological principles.

Although computers will become more important and conspicuous in all areas of biology it is difficult to predict the areas in which they will have their greatest impact. Hardware and software improvements will produce ever more powerful computer systems which may allow the development of presently unrealised applications. In addition, future generations of biologists will probably be more willing (and able?) to use computers. However, biologists will need to expand their links with workers from other disciplines, particularly mathematicians and physicists, if they are to exploit the enormous potential of computers in biology. Nonetheless, I would expect that at least five areas of computer applications will become almost routine in biology in the near future.

If we exclude word processing, the most commonplace application will continue to be number crunching, almost invariably in the form of statistical analyses. If this is carried out on a small scale then BASIC will provide more than adequate programs. However, if the analysis is concerned with large bodies of data or multivariate techniques which require the application of matrix algebra, BASIC will need to be replaced by a more powerful programming language. Although it may be possible to write such programs in BASIC they will be very slow to execute. This does not mean that small microcomputers have no part to play in large number crunching applications. Assuming that the relevant communications software and hardware is available, microcomputers can be used for the collation and preliminary processing of large data sets using programs written in BASIC. The output from these preliminary operations could be written to a disk file in an ASCII format, i.e. each character is stored as its ASCII equivalent. This would mean that a number such as 12.345 would not be stored in floating point notation but as a string of characters '1' '2' '.' '3' '4' '5'. Data stored in this format can be transferred between most computers and programming languages.

Equipment interfacing, and the related field of remote data logging, will continue to grow in importance. Data logging is the technique of using a microcomputer to collect batches of data from a piece of equipment. It can be particularly important in field experiments where it is not possible or convenient for the experimenter to be present. BASIC may be an adequate programming language for writing interfacing software on computers such as the BBC microcomputer (which has interfacing commands built into the BASIC dialect) but for the majority of computers it may be necessary to write the interfacing software in machine language or an efficient language such as FORTH. The alternative is to purchase an interfacing system which contains both the hardware and software.

The third of the applications will be modelling. Because most complex biological problems tend to be multivariate any realistic models of these systems must also be complex. Inevitably such models can only be implemented on computers. An interesting feature of the increasing power of microcomputers is that even relatively complex models will become available to a wider audience. The scope of computer models covers an enormous range but includes areas such as epidemiology, prediction of molecular structures and the dynamics of ecological systems.

The old proverb that 'a picture paints a thousand words' could perhaps be amended to 'a picture paints a thousand numbers'. Computer graphics provides a method for testing this proverb. The power and speed of microcomputer graphics systems has been increasing rapidly. Computer graphics can be used to visualise mathematical relationships and to provide a medium with which more biologists are happy. Apart from the graphical presentation of data, the two most important applications will probably be in the areas of molecular biology and image analysis techniques applied to microscopic structures.

The fifth of the potential 'commonplace' applications will be in the area of databases and expert systems. Most current research into database systems is directed at commercial applications, nonetheless the techniques developed should be equally relevant in the biological sphere. One of the most rapidly growing biological applications is that of DNA databases. DNA base sequences are particularly suitable for inclusion in a database system because of the very large number of base sequences now available and the need to be able to compare similar base sequences from a variety of sources.

Expert systems are, to a certain extent, extensions of the database concept combined with ideas from artificial intelligence research. The number of successful expert systems is beginning to increase as the background theory and necessary algorithms are further developed. An expert system is a program which can operate in a very restricted area and which can provide opinions based upon facts which have been presented to the program. The program reaches its conclusions by applying rules ('knowledge') which have been supplied by human experts in the field. These rules will take the form of structures such as 'if a is true then consider option b'. They are not based on strict logic but follow the technique and 'rules' used by a human expert. Often the most difficult part is identifying and formalising the expert's rules. One of the existing expert systems which may be of interest to the biologist is MYCIN, which is used to diagnose blood infections. It is provided with a set of over 500 rules and follows a procedure which includes determining whether the organism is gram positive or negative, its morphology and colony characteristics. When it has obtained answers to these and any subsequent questions the program will suggest a probable identity and potential course of treatment for the infection. Expert systems such as MYCIN and DENDRAL (a system used to determine the structure of organic molecules) are essentially diagnostic. Much of the current research into expert systems is directed at producing systems which can be used to discover new information or provide additional proofs for existing information. There are many biological and medical applications in which expert systems could be of enormous value.

After reading these five applications the reader may be wondering why they should learn BASIC. There are several reasons, none of which excludes the possibility of learning an alternative programming language if the reader feels capable. BASIC is a very accessible and forgiving programming language. It is implemented, almost universally, on microcomputers as an interpreted language and this allows for a large degree of flexibility when experimenting with programs. Because of these reasons BASIC allows a newcomer to become familiar with many aspects of computer science which previously appeared to be incomprehensible. As long as the user thinks about the structure of the programs and follows, whenever possible, the 'rules' of structured programming, it should be possible to transfer to a structured language, such as Pascal, without too much difficulty.

It is possible that at some time in the future the biologist may be able to make a contribution to computer science. Although it is currently in the realms of science fiction there may come a time when microprocessors are replaced by 'biochips'. There is a finite limit to the miniaturisation of electronic circuits. Even present day integrated circuits are approaching the size at which damage from cosmic rays and other types of ionising radiation can be a problem. Perhaps another approach is to make use of the information handling capabilities of biological molecules such as proteins and nucleic acids?

Appendix A Number Storage Systems and Sources of Error

A.1 Integer storage and arithmetic

As described in Chapter 1 a byte can be used be used to store an unsigned integer between 0 and 255 (binary 00000000 to 11111111). However, this arrangement vastly reduces the capabilities of computers. If signed (positive and negative) integers are to be stored one of the bits must be used to represent the sign. Bit 7, the leftmost bit, is used to signify the sign; 0 indicates a positive number; while 1 is used with negative numbers. Since bit 7 represents the sign it is said to be the most significant bit. Only 7 bits (bits 0 to 6) are now available to store the number. Thus a single byte can be used to store a signed integer between −127 and +127. However, although +127 would be stored as 01111111, −127 would not be stored as 11111111. This is because binary arithmetic will not always work correctly if this format is used. (Note that binary arithmetic follows similar rules to decimal arithmetic.)

Consider the following examples:

	Decimal	Binary	
a)	8 +	00001000 +	
	3	00000011	
	11	00001011	correct
b)	8 +	00001000 +	
	−3	10000011	
	+5	10001011	incorrect

Example (b) is incorrect because 10001011 would be −11. Binary arithmetic will only work correctly if negative numbers are stored in a 'two's complement' format. This format is derived by swapping all of the bits, except bit 7. Thus all of the ones become zeros and vice versa. Finally one is added to the result. Thus:

unsigned decimal 3 is binary	00000011
signed decimal −3 is binary	10000011
complement of signed decimal −3 is binary	11111100
two's complement of signed decimal −3 is binary	11111101

If the 8 + (−3) addition is repeated but using two's complement notation the correct result will be obtained.

```
Decimal  Binary
   8 +   00001000 +
  -3     11111101
  ----   --------
  +5     00000101
```

The byte 00000101 is decimal +5 which is now correct. Note that when two binary ones are added a carry is produced. In this case the carry is eventually lost from the leftmost end of the byte. This is an external carry.

If the answer is negative the decimal equivalent can be obtained by taking the two's complement of the result. Thus:

```
Decimal  Binary
  -8 +   11111000 +
   3     00000011
  ----   --------
  -5     11111011
```

The complement of bits 0 to 6 of the byte 11111011 is 10000100. Adding one gives the two's complement 10000101 which is −5, the correct result. Problems still occur if one of the following conditions arises:

i) the result is greater than +127 or less than −127;
ii) a large positive number is subtracted from a large negative number;
iii) a large negative number is subtracted from a large positive number.

For example:

```
Decimal  Binary
  70 +   01000110 +
  64     01000000
  ----   --------
 +134    10000110
```

The status of bit 7 indicates a negative result! If the two's complement of 10000110 is taken a decimal answer of −122 (11111010) is obtained. The problem is due to an internal carry from bit 6 to bit 7 which has changed the sign. Most CPUs have a special register called the status register. The status register contains a byte whose bits are set (changed from 0 to 1) whenever certain conditions arise. Each of the bits in the status register is said to be a flag. Two of these flags should be consulted whenever binary arithmetic is carried out. The flags concerned are the overflow and carry flags. The overflow flag is set if there is an internal carry from bit 6 to bit 7, while the carry flag is set if there is an external carry.

An incorrect result will be obtained when:

i) there is a carry from bit 6 to 7 but no external carry
ii) there is an external but no internal carry.

These conditions can be detected by consulting the overflow and carry flags (there are machine language instructions which perform these operations). Once the error has been detected appropriate actions can be taken to obtain the correct result.

Integer numbers greater than 127 can be stored if two or more bytes are used to store a number. In most microcomputers a maximum of two bytes is used to store integers. This gives a range of +32767 to −32767. The bits are numbered 0 to 15 which is equivalent to 2^0–2^{15}. Bits 0 to 7 are in the righthand byte while bits 8 to 15 are in the leftmost byte. Bit 15, the leftmost bit, is the sign bit. Since the leftmost byte carries the sign bit it is said to be the most significant byte (MSB) while the righthand byte is the least significant byte (LSB). The contents of these two bytes can be easily calculated without resorting to powers of 2. The MSB will have a value of the modulus of number/256, while the LSB contains the remainder. Thus +20354 would be stored as:

MSB = modulus 20354/256 = 79 (01001111)
LSB = (20354 − (256 × 79)) = 130 (10000010)

Note that bit 7 in the LSB does not carry the sign. An error will be produced if the result of any integer arithmetic is greater than +32767 or less than −32767.

A.2 Real numbers

Real numbers, that is those which can take any value, present a different problem. Examples of real numbers would be 0.00234, 3.1416, 3241100023.0. In order to conserve memory and ease computations each of these numbers must be stored in a common format. This format, which is similar to the familiar scientific notation, is called floating point notation. Any real number can be described by three characteristics:

a) its sign
b) its significant digits
c) its magnitude.

In order to achieve a satisfactory numeric range and level of precision each real number must be stored in a group of at least four bytes.

Before a real number can be stored it must first be normalised. This involves converting a number into the following format:

$$M \times base^E$$

M is the mantissa and E the exponent. Thus 0.00234 could be written as 2.34×10^{-3}, while 32,411,000 would be 3.2411×10^7. When a number is stored in a floating point format one byte is used to store the exponent and three bytes are used to store the mantissa. Since the mantissa must be stored as a signed number the first bit, in the 3 bytes, is used to store the sign. The remaining 23 bits hold the digits of the mantissa. This places a constraint on the number of mantissa digits which can be stored. Typically this is 7 to 8 decimal digits.

The actual storage systems used will vary with the type of computer but it is very rare for numbers to be stored with a base of 10. The base used will normally be 2, 8 or 16, for example $M \times 2^E$. This is harder for humans to understand but simplifies the design and operation of computers. The exponent may be stored in an 'excess' format. If this system is used an exponent within the range -64 to $+64$ can be stored. The excess format works by adding 64 to the exponent. Thus, -64 becomes 0 and $+64$ becomes 128. This has several advantages. It avoids the need for an exponent sign bit and allows computations to proceed faster. In most 8-bit microcomputers the floating point storage system provides a normal range of -1×10^{38} to 1×10^{38}.

A.3 Sources of error

Most people assume that computers always produce accurate results from computations but there will be occasions when this not true. Two general categories of error can be recognised: 'computer' errors and human errors. The computer errors almost always result from the numeric storage systems used.

A.3.1 Conversion errors

Data is usually presented to the computer in a fixed point format, for example: 1.2, 326.25, 0.00345. Before this data can be used by the computer it must be converted into a floating point format. This process involves normalisation and base conversion. Whenever these conversions (and conversions back to base 10 numbers) occur a small error (loss of accuracy) may be introduced. The most severe errors are associated with calculations involving a combination of very small and very large numbers (as in the energy balance model described in Chapter 11).

A.3.2 Rounding errors

Computers can produce rounding errors while carrying out arithmetic, storing numbers and outputting numbers. Integers, in particular, must be treated with caution: integer arithmetic may produce some surprising results, for example $11/4 = 2$! Also, the INT function does not carry out a true rounding on real numbers. Any integer calculations which produce a result outside the integer storage range will also force an error. Even when real numbers are used there will be errors (in addition to those described in A.3.1). These result mainly from the level of precision operating which is determined by the number of bits available for storage of the mantissa. If 7 digit precision is allowed a number such as 10.233040778 would become 1.023304×10^2. The last 4 digits will be lost. Finally a number which is within the range $+/- 2.9388 \times 10^{-39}$ may be rounded off to zero. This can cause additional problems since it could produce a 'division by zero' error if the denominator in a division is within this range. Some computers allow greater levels of precision to be specified. This is achieved by allocating more bytes for storage of the mantissa.

A.3.3 BASIC function errors

Many of the BASIC functions, particularly the trigonometric functions, are evaluated by using algorithms which produce approximations to the actual value. For example, the sine of an angle is evaluated from the summation of a series. The interpreter designer must, however, strike a compromise between speed and accuracy. Consequently, SIN(X) is evaluated from a limited number of terms within the series. This produces a function which is an approximation to the correct value.

A.3.4 Human errors

Humans are capable of generating many computer errors! Probably the most difficult ones to detect are those which result from algorithm or logic errors. It is good practice, during program development, to run the program with some test data which can be used to verify that the program is working correctly. If this data is held in DATA statements, during program development, the program can be tested many times without the need to re-enter any data. Obviously the data input statements will need to be amended to their final format before the program is finished.

There are a number of common errors which are made during data entry. These can only be detected by careful data checking. The two most common errors are:

    i) lost digit — for example 888783 entered as 88783
   ii) transposition — for example 545 entered as 454.

Appendix B ASCII Conversion Table

Decimal	Hex	Character	Decimal	Hex	Character	Decimal	Hex	Character
0	00	NUL	43	2B	+	86	56	V
1	01	SOH	44	2C	,	87	57	W
2	02	STX	45	2D	−	88	58	X
3	03	ETX	46	2E	.	89	59	Y
4	04	EOT	47	2F	/	90	5A	Z
5	05	ENQ	48	30	0	91	5B	[
6	06	ACK	49	31	1	92	5C	\
7	07	BEL	50	32	2	93	5D	]
8	08	BS	51	33	3	94	5E	^
9	09	HT	52	34	4	95	5F	←
10	0A	LF	53	35	5	96	60	
11	0B	VT	54	36	6	97	61	a
12	0C	FF	55	37	7	98	62	b
13	0D	CR	56	38	8	99	63	c
14	0E	SO	57	39	9	100	64	d
15	0F	SI	58	3A	:	101	65	e
16	10	DLE	59	3B	;	102	66	f
17	11	DC1	60	3C	<	103	67	g
18	12	DC2	61	3D	=	104	68	h
19	13	DC3	62	3E	>	105	69	i
20	14	DC4	63	3F	?	106	6A	j
21	15	NAK	64	40	@	107	6B	k
22	16	SYN	65	41	A	108	6C	l
23	17	ETB	66	42	B	109	6D	m
24	18	CAN	67	43	C	110	6E	n
25	19	EM	68	44	D	111	6F	o
26	1A	SUB	69	45	E	112	70	p
27	1B	ESC	70	46	F	113	71	q
28	1C	FS	71	47	G	114	72	r
29	1D	GS	72	48	H	115	73	s
30	1E	RS	73	49	I	116	74	t
31	IF	US	74	4A	J	117	75	u
32	20	SP	75	4B	K	118	76	v
33	21	!	76	4C	L	119	77	w
34	22	"	77	4D	M	120	78	x
35	23	#	78	4E	N	121	79	y
36	24	$	79	4F	O	122	7A	z
37	25	%	80	50	P	123	7B	{
38	26	&	81	51	Q	124	7C		
39	27	'	82	52	R	125	7D	}
40	28	(	83	53	S	126	7E	ˉ
41	29	)	84	54	T	127	7F	DEL
42	2A	*	85	55	U			

Hex = Hexadecimal

The first 32 characters are control codes. Many of these refer to data transfer operations and paper movements which were found on the early teletype terminals. Table B.1 lists the actions of these control codes.

Table B.1 Actions of control codes

Decimal ASCII code	Abbreviation	Action
0	NUL	Nothing
1	SOH	Start of heading
2	STX	Start of text
3	ETX	End of text
4	EOT	End of transmission
5	ENQ	Enquiry
6	ACK	Acknowledge
7	BEL	Bell
8	BS	Backspace
9	HT	Horizontal tabulation
10	LF	Linefeed
11	VT	Vertical tabulation
12	FF	Form feed
13	CR	Carriage return
14	SO	Shift out
15	SI	Shift in
16	DLE	Data link escape
17	DC1	Device controls
18	DC2	Device controls
19	DC3	Device controls
20	DC4	Device controls
21	NAK	Negative acknowledge
22	SYN	Synchronous idle
23	ETB	End of transmission block
24	CAN	Cancel
25	EM	End of medium
26	SUB	Substitute
27	ESC	Escape
28	FS	File separator
29	GS	Group separator
30	RS	Record separator
31	US	Unit separator
32	SP	Space

Microcomputer Interfacing

This discussion is restricted to 8-bit microcomputers.

C.1 Introduction

Now that inexpensive microcomputers are readily available they are being linked to a wide range of scientific, laboratory and field equipment. These links allow the user to establish computer control over equipment and data collection from instruments (sensors). However, before a link can be established a suitable interface, between the computer and a piece of external equipment, must be provided. An interface is a device which allows peripheral equipment to communicate with the CPU. All computers will contain a number of existing interfaces which link the CPU with its input, output and storage devices. Whenever a new peripheral, for example, a printer or a piece of scientific equipment, is added to a computer system an additional interface must be provided. Successful interfacing depends not only upon the design of the interface hardware, but also on the software which will exploit the interface and its peripheral. Thus it is possible to talk about hardware and software interfacing.

Since an interface is concerned with the management of communications it must be capable of controlling information input and output (I/O). It is important, therefore, to understand the communication channels which are available within a computer system. Information is transported around the computer system using parallel circuits called buses of which there are three types:

a) the address bus
b) the data bus
c) the control bus.

The address bus is used to transport memory addresses. It can be thought of as 16 wires, or lines, labelled A0 to A15. Each line will carry a voltage of either 0 or 5 volts. These are respectively equivalent to logical 0 and logical 1. Thus, by applying 5 volts to the correct combination of 16 lines an address within the range 0 to 65535 can be generated. (16-bit microprocessors usually have an address bus composed of more lines and can, therefore, address more memory locations.) Special address decoding circuits can be constructed which use TTL devices. A TTL device (transistor –

transistor logic) is an integrated circuit composed of several transistors formed on a single layer of silicon. These circuits can recognise which address is currently on the address bus by decoding the bit-voltage patterns.

The data bus consists of 8 lines, labelled D0 to D7. Each of these lines will again carry either 0 or 5 volts. Thus, if the lines D0, D1 and D3 carried 5 volts the data on the data bus would be 00001011 in binary, or decimal 11. The CPU 'writes' information onto the data bus by generating the appropriate signals. The CPU 'reads' information from the data bus when the voltages on the data bus were generated by one of the peripherals.

The control bus is made up of a number of lines, normally about 6, which carry command signals to and from the peripherals. For example, depending upon the CPU, one or two of the control lines will be used to inform a peripheral when the CPU requires to read or write information. Other important control lines are associated with interrupts. A peripheral can generate an interrupt signal by setting the interrupt line to the appropriate voltage. When the CPU receives an interrupt it has to abort the work which it is currently undertaking and establish active communication with the peripheral. However, this must be done in such a way that none of the existing information is corrupted or destroyed. Therefore, before responding to the interrupt the CPU saves the current contents of its internal registers for later retrieval when the interrupt has been processed. Interrupts can be very important when peripherals may have data available at unpredictable times.

There are occasions when a peripheral generates data faster than the CPU can normally process it. Loss of data can be avoided in these circumstances if Direct Memory Addressing (DMA) is used. The CPU is then bypassed and the peripheral writes its information directly into the memory, from here it can be processed by the CPU when time is not critical. Obviously controls must be established to ensure that the peripheral does not overwrite areas of memory which are currently in use.

When a peripheral is interfaced it must be connected, directly or indirectly, to at least one of the buses. This means that each peripheral must have a mechanism for detecting when the CPU is attempting to establish communications with it rather than with any of the other peripherals. Two main systems are used. These are the 'memory mapped' and 'port based' systems. The two systems result from differences in CPU design and construction. If a memory mapped system is used each peripheral is allocated a portion of memory (as in the memory mapped screens described in Chapter 9.3). The peripheral interfaces have address decoders which activate the interface when the relevant memory address is placed on the address bus. Some manufacturers make provisions for extra peripherals when designing their computers. Thus, the Apple II microcomputer has eight 'slots' which can be used to connect peripherals. Each slot is connected to all three buses and is allocated a unique block of 16 bytes in RAM. In addition each slot is provided with an address decoder which detects when the slot is 'active'. This system allows Apple II computers to be interfaced to a wide range of peripherals in a relatively simple manner.

Microprocessors such as the Z-80 are provided with I/O ports. The processor can be used to communicate directly with 256 different devices using a number of special machine instructions. This system is useful because it simplifies the design and construction of interfaces. Most BASIC dialects, implemented on Z-80 microcomputers, provide high level commands such as OUT and IN which can be used to communicate with external devices. Z-80 based microcomputers can also use the memory mapped system for interfacing.

C.2 Analogue devices

Most of the equipment interfaced, or used, by biologists will generate an analogue output. An analogue signal produced by scientific equipment will normally be a continuously varying voltage or current. Analogue signals cannot be directly processed by computers but must be converted to a digital format. This conversion will be carried out by an Analogue to Digital (A/D) converter. These can be purchased for most microcomputers. It is also possible to use Digital to Analogue (D/A) converters. These produce an analogue output which can be used to control external equipment (such as a chart recorder). All A/D converters place a restriction on the range of acceptable analogue input voltages. Often these ranges will not match up with those produced by the equipment which you are attempting to interface to the computer. Consequently, the input signal must be adjusted by using a signal conditioning unit (a relatively simple electronic device which can be used to amplify or attenuate the output signal). Even when the two ranges are compatible it is good practice to continue using the conditioning unit. This is because electronic equipment often generates a voltage outside its nominal range during switching operations. A voltage surge which was fed directly into the computer could cause serious damage.

Many microcomputers are provided with joystick inputs for computer games. These inputs are in fact relatively simple A/D converters. A joystick contains two potentiometers which produce output voltages depending upon the position of the stick. The two analogue voltage inputs are converted into two 8-bit binary numbers (decimal 0 to 255) by the converter. An analogue voltage input from any source, assuming that it falls within the correct limits for the A/D converter, could be utilised by these converters.

An A/D converter works by breaking down an input voltage range into a number of discrete steps. The number of steps, which determines the resolution of the converter, depends upon the output word size. An 8-bit output word allows 256 incremental steps. The converter works by generating an internal voltage which depends upon the value of a binary word. This internal voltage is compared with the input voltage and adjusted up or down as appropriate. When the two match, within the limits of resolution, the binary word is output. The degree of resolution which would be acceptable depends upon many factors. Consider a device which generates an analogue output voltage with a full scale range of 0 to 10 volts. If an A/D converter with a 4-bit word is used this would provide a resolution of 16 incremental steps of 0.625 volts. An 8-bit output word increases the resolution to 256 steps of 0.0391 volts. 16-bit resolution would provide 65536 steps of 0.00015 volts. A 4-bit word would provide a resolution which is too coarse for most biological applications, while a 16-bit word may provide a resolution greater than the accuracy of the equipment producing the analogue output. This would result in a large amount of 'noise' in the input data. A reasonable compromise is provided by a converter using a 12-bit word. This provides 4096 incremental steps, of 0.0024 volts in our example.

Other factors which would need to be considered when selecting a converter include the response time and the amount of memory required. An A/D converter will take a finite time to process a signal and it cannot accept the next signal until the existing one has been dealt with. The speed at which information is processed will determine the sampling rate of the converter. If a very high rate of data acquisition is required it is likely that an expensive converter will be needed. There is a relationship between the sampling rate and the amount of memory required. A high speed of data acquisition can quickly fill the available memory. When designing or selecting an interfacing system constraints such as these need to be considered and the end result will inevitably be a compromise between speed, precision and cost.

Before attempting any interfacing it is strongly recommended that the user consults specialist texts on this subject. Books are available for most of the common microcomputers. A more detailed introduction to A/D interfacing can be found in Geisow and Barrett's book (see Appendix E).

Answers to selected problems

```
1  REM            PROBLEM 3.1
10 REM            WET WEIGHT TO DRY WEIGHT CONVERSION
11 REM            W IS THE WET WEIGHT
12 REM            D IS THE DRY WEIGHT
13 REM
20 PRINT "ENTER THE WET WEIGHT ( IN G )"
30 INPUT W
40 LET D = W * .653
50 PRINT "THE DRY WEIGHT IS"
60 PRINT D
70 END
```

```
1  REM            PROBLEM 3.2
10 REM            ESTIMATION OF CARDIAC OUTPUT
11 REM  VARIABLES USED
12 REM            O - OXYGEN CONSUMPTION IN ML / MIN
13 REM            A - ARTERIAL OXYGEN CONC IN ML / L
14 REM            V - VENOUS OXYGEN CONC IN ML / L
15 REM            02 - CARDIAC OUTPUT IN L / MIN
16 REM
20 PRINT "ENTER THE FOLLOWING "
30 PRINT "  OXYGEN CONSUMPTION ( ML / MIN )"
35 INPUT O
40 PRINT "  ARTERIAL OXYGEN CONCENTRATION ( ML / L)"
45 INPUT A
50 PRINT "  VENOUS OXYGEN CONCENTRATION ( ML / L)"
55 INPUT V
60 LET 02 = O / ( A - V )
70 PRINT
75 PRINT "CARDIAC OUTPUT ( L / MIN ) IS "
80 PRINT 02
90 END
```

```
1   REM             PROBLEM 3.3
10  REM             ESTIMATION OF BODY SURFACE AREA
11  REM
12  REM             W - BODY WEIGHT IN KG
13  REM             H - HEIGHT IN CM
14  REM             S - SURFACE AREA IN SQUARE METERS
15  REM
20  PRINT "ENTER THE BODY WEIGHT IN KG "
25  INPUT W
30  PRINT "ENTER THE HEIGHT IN CM "
35  INPUT H
40  LET S = .007184 * W^.452 * H^.725
45  PRINT
50  PRINT "BODY SURFACE AREA IN SQUARE METERS IS "
60  PRINT S
70  END


1   REM             PROBLEM 3.4
10  REM             DETERMINATION OF INFRA RED RADIATION
11  REM             FROM THE SKY
12  REM
13  REM             T - TEMPERATURE IN CELCIUS
14  REM             E - SATURATION VAPOUR PRESSURE IN MBAR
15  REM             I - INFRA RED RADIATION WATTS / SQUARE METER
16  REM             S - STEFANS CONSTANT  0.000000557
17  REM
18  LET S =  5.57E-07
20  PRINT "ENTER THE AIR TEMPERATURE IN DEGREES CELCIUS"
25  INPUT T
26  REM             CONVERT TEMPERATURE TO KELVIN
27  LET T = T + 273
30  PRINT "ENTER THE SATURATION VAPOUR PRESSURE IN MBAR"
35  INPUT E
40  PRINT
50  LET I = ( .04 + .06 * E^.5 ) * S * T^4
60  PRINT "INFRA RED RADIATION IN WATTS / SQUARE METER IS "
70  PRINT I
80  END
```

```
10 REM            PROBLEM 4.1
20 REM            SUM OF A LIST OF NUMBERS
30 REM            N — NUMBER OF ELEMENTS IN THE LIST
40 REM            X( ) — AN ARRAY USED TO STORE THE NUMBERS
45 REM            RESERVE ENOUGH SPACE FOR 50 NUMBERS
50 REM            T — TOTAL
55 REM            INITIALISATION
60 DIM X ( 50 )
70 LET N = 0
80 LET T = 0
90 REM            READ IN THE LENGTH OF THE LIST
100 READ N
110 REM            READ IN THE NUMBERS
120 FOR I = 1 TO N
130 READ X ( I )
140 LET T = T + X ( I )
150 NET I
160 PRINT "TOTAL IS "; T
170 END
179 REM            FIRST DATA VALUE IS N
180 DATA 5
190 DATA 66,87,123,24,35


170 REM            PROBLEM 4.2
171 REM            APPEND TO THE END OF PROBLEM 4.1
172 REM            REMEMBER TO REENTER THE DATA STATEMENTS
174 REM            SELECTION OF AN ARRAY ELEMENT
175 REM
180 PRINT "ENTER THE NUMBER OF THE ARRAY ELEMENT REQUIRED ";
185 INPUT Q
190 PRINT "PRESENT VALUE IS "; X ( Q )
200 END


170 REM            PROBLEM 4.3
171 REM            APPEND TO THE END OF PROBLEM 4.1
172 REM            REMEMBER TO REENTER THE DATA STATEMENTS
174 REM            CHANGE AN ELEMENT IN THE ARRAY X( )
175 REM
180 PRINT "ENTER THE NUMBER OF THE ARRAY ELEMENT REQUIRED ";
185 INPUT Q
190 PRINT "PRESENT VALUE IS "; X ( Q )
200 PRINT "WHAT IS THE NEW VALUE ";
210 INPUT X ( Q )
220 REM            PRINT OUT THE NEW ARRAY
230 FOR I = 1 TO N
240 PRINT X ( I )
250 NEXT I
270 END
```

```
100 REM          PROBLEM 5.1
101 REM          A FUNCTION TO EVALUATE LOG 10 OF A NUMBER
102 REM
110 DEF FNA ( X ) = LOG ( X ) / 2.3026
120 REM          THE FUNCTION CAN NOW BE USED



1   REM          PROBLEM 5.2
10  REM          HOW DOES DEPTH AFFECT LIGHT INTENSITY?
20  PRINT "CALCULATION OF LIGHT INTENSITY AT VARIOUS"
21  PRINT "DEPTHS IN WATER"
22  PRINT
30  PRINT "ENTER THE LIGHT INTENSITY AT THE SURFACE ";
35  INPUT I0
40  PRINT "ENTER THE APPROPRIATE EXTINCTION COEFFICIENT ";
45  INPUT K
50  PRINT "ENTER THE DEPTH ";
55  INPUT Z
60  PRINT
70  I2 = I0 * EXP ( - ( K * Z ) )
80  PRINT "LIGHT INTENSITY AT DEPTH ";Z;" IS ";I2
90  END



1   REM          PROBLEM 5.3
10  REM          ESTIMATION OF NET ASSIMILATION OF A CROP
11  REM     VARIABLES USED
12  REM          T - TIME PERIOD (NO OF DAYS, WEEKS ETC )
13  REM          W1, W2 THE WEIGHTS OF THE PLANTS
14  REM          L1, L2 THE LEAF AREAS OF THE PLANTS
15  REM          NET ASSIMILATION RATE G / SQ MM / UNIT TIME
20  PRINT "THIS PROGRAM ESTIMATES THE NET ASSIMILATION OF A"
21  PRINT "CROP USING LEAF AREA AND WEIGHT DATA OBTAINED AT THE"
22  PRINT "BEGINNING AND END OF A SPECIFIED PERIOD OF TIME"
25  PRINT
30  PRINT "ENTER THE FOLLOWING INFORMATION"
40  PRINT "   TIME PERIOD ";
45  INPUT T
50  PRINT "   WEIGHT AT THE BEGINNING ";
55  INPUT W1
60  PRINT "   WEIGHT AT THE END OF THE PERIOD ";
65  INPUT W2
70  PRINT "   LEAF AREA AT THE BEGINNING ";
75  INPUT L1
80  PRINT "   LEAF AREA AT THE END OF THE PERIOD ";
85  INPUT L2
90  LET E = ( W2 - W1 ) / T * ( LOG (L2) - LOG (L1) ) / ( L2 - L1 )
95  PRINT
100 PRINT "NET ASSIMILATION = ";
105 PRINT E
110 END
```

```
201 REM         PROBLEM 6.1
202 REM         ASSUME BLOOD PRESSURES ARE STORED IN ARRAY P( )
203 REM         NUMBER OF PATIENTS IS N
204 REM
210 PRINT "THE FOLLOWING PATIENTS HAVE A SYSTOLIC BLOOD"
211 PRINT "OF GREATER THAN 125 MM HG"
220 FOR I = 1 TO N
230 IF P ( I ) > 125 THEN PRINT I
240 NEXT I
250 REM         PROGRAM WOULD CONTINUE FROM HERE


201 REM         PROBLEM 6.2
202 REM         ASSUME BLOOD PRESSURES ARE STORED IN ARRAY P( )
203 REM             BODY WEIGHTS ARE IN AN ARRAY W( )
204 REM         THERE ARE N PATIENTS
205 REM
210 PRINT "THE FOLLOWING PATIENTS HAVE A SYSTOLIC BLOOD PRESSURE"
211 PRINT "GREATER THAN 115 MM HG AND WEIGH OVER 85 KG"
220 FOR I = 1 TO N
230 IF P ( I ) > 115 AND W ( I ) > 85 THEN PRINT I
240 NEXT I
250 REM         PROGRAM WOULD CONTINUE FROM HERE


1   REM     PROBLEM 7.5
10  REM     COUNT THE AMINO ACID FREQUENCIES IN A POLYPEPTIDE
20  REM     USING THE SINGLE LETTER CODE
30  REM     STORE THE FREQUENCIES IN AN ARRAY C( ). MUCH OF
40  REM     THE PROGRAM IS THE SAME AS THE PROGRAM IN 7.4.2
50  REM     INITIALISE THE VARIABLES
51  DIM C(26)
52  LET X = 0
53  LET S$ = ""
60  READ S$
70  FOR I = 1 TO LEN ( S$ )
80  LET  X = ASC ( MID$ ( S$, I, 1 ) ) - 64
82  REM       SKIP ANY NON LETTER ENTRIES
84  IF X < 1 OR X > 26 THEN GOTO 100
90  LET C ( X ) = C ( X ) + 1
100 NEXT I
110 REM
120 REM      NOW PRINT OUT THE FREQUENCIES
130 REM
140 FOR I = 1 TO 26
150 REM      SKIP ASCII CODES FOR J, O, U & X
155 LET I2 = I + 64
160 IF I2 = 74 OR I2 = 79 OR I2 = 85 OR I2 = 88 THEN GOTO 180
170 PRINT CHR$ ( I2 ), C ( I )
180 NEXT I
189 REM      TEST DATA HUMAN ALPHA HAEMOGLOBIN ( 1ST 40 AMINO ACIDS )
190 DATA V LSPADKTNVKAAWGKVGAHAGEYGAEALERMFLSFPTT
```

Appendix E Bibliography

Chapter 1
Glinert, E. (1983) *Introduction to Computer Science using Pascal.* Prentice-Hall International, Englewood Cliffs, New Jersey.
Giesow, M. J. and Barrett, A. N. (1983) *Computing in the Biological Sciences.* Elsevier Biomedical Press, Amsterdam.
Goldschlager, L. and Lister, A. (1982) *Computer Science: A Modern Introduction.* Prentice-Hall International, Englewood Cliffs, New Jersey.
Walker, R. S. (1981) *Understanding Computer Science.* Texas Instruments Learning Center, Dallas.
There is a new journal, *Computer Applications in the Biosciences,* which is published by IRL press. The pilot issue was published in June 1984.

Chapter 2
No specific BASIC programming books are recommended. This is because of the problems resulting from the many differences between BASIC dialects.
Findlay, W. and Watt, D. A. (1981) *Pascal An Introduction to Methodological Programming 2nd Edition.* Pitman, Bath.
Munro, D. M. (1982) *FORTRAN 77.* Edward Arnold, London.
Spence, J. W. (1982) *COBOL for the 80's.* West Publishing Co., St. Paul.
Winfield, A. (1983) *The Complete Forth.* Sigma Technical Press, Wilmslow, UK.

Chapter 4
Cooke, D., Craven, A. H. and Clarke, G. M. (1982) *BASIC Statistical Computing.* Edward Arnold, London.
Rogers, D. W. (1983) *BASIC Microcomputing and Biostatistics.* Humana Press Inc., Clifton, New Jersey.
Zar, J. H. (1974) *Biostatistical Analysis.* Prentice-Hall International, Englewood Cliffs, New Jersey.

Chapter 8
Bourne, S. R. (1983) *The UNIX System.* Addison-Wesley, Wokingham, UK.
Dietel, H. M. (1983) *An Introduction to Operating Systems.* Addison-Wesley, Wokingham, UK.
Hogan, T. (1982) *Osborne CP/M User Guide.* Osborne/McGraw-Hill, Berkeley.

Chapter 9
Artwick, B. A. (1984) *Applied Concepts in Microcomputer Graphics.* Prentice-Hall, Inc., Englewood Cliffs, New Jersey.
Foley, J. and van Dam, A. (1982) *Fundamentals of Interactive Computer Graphics.* Addison-Wesley, Wokingham, UK.

Chapter 11

Gates, D. M. (1972) *Man and his Environment: Climate.* Harper and Row, New York.

Pielou, E. C. (1974) *Population and Community Ecology: Principles and Methods.* Gordon and Breach, New York.

Spain, J. D. (1982) *BASIC Microcomputer Models in Biology.* Addison-Wesley, Wokingham, UK.

Chapter 12

Date, C. J. (1981) *An Introduction to Database Systems Volume I.* Addison-Wesley, Wokingham, UK.

Date, C. J. (1982) *An Introduction to Database Systems Volume II.* Addison-Wesley, Wokingham, UK.

Meadows, A. J., Gordon, M. and Singleton, A. (1982) *Dictionary of New Information Technology.* Century, London.

Chapter 13

James, M. (1982) *The 6809 Companion.* Bernard Babini, London.

Norusis, M. J. (1982) *SPSS Introductory Guide: Basic Statistics and Operations.* McGraw-Hill, New York.

Ryan, Jr., T. A., Joiner, B. L. and Ryan, B. F. (1976) *MINITAB Student Handbook.* Duxbury Press, Massachusetts.

Index

A/D conversion   178–179
ABS   52, 62
address bus   7, 8, 176–177
Agris   147
algorithm   13, 20–21, 36, 44–45, 57,
    95–96, 162
ALU   7, 9
amino acid sequence   84–86
analysis of variance   47
AND   73
ANSI   9
ARCSIN   54
argument   52
arithmetic operators   28–29, 33
ASC   78, 87
ASCII   9, 77, 81–82, 140
ASCII file   166
assembly language   16
assignment   25
asynchronous transmission   141
ATN   52, 54, 61

backing store   10–11
band width   104
baseband transmission   142
BASIC   14, 16–17, 20
batch processing   153, 157
baud rate   141
benchmark   153
binary arithmetic   169–171
binary file   99
binary number system   4, 169–171
Bio Research Index   147
Biological Abstracts   147
Biosis Previews   147
bit   4
BLAISE   147
broadband transmission   142
broadcast videotext   146
bubble sort   119–120
byte   4

CAI   2
Cambridge ring   144

Cancerline   147
Cartesian coordinate   106–108
CEEFAX   146
Centronics interface   140
checksum   141
Chemdex   147
CHR$ 78–79, 87
chromosome   111–112
CLOSE   97
clock   7, 9
clock signal   141
CLS   43
COBOL   18
coding   13, 14, 20
comma   51
competition   134–138
compiler   17, 161–162
complement   169–170
concatenation   78
concurrency   100
conditional command   64, 67–76
control bus   176–177
control character   141
control code   79, 175
control key   9
control unit (CPU)   7
conversion error   172
COS   52, 54, 61
COSECANT   54
COTANGENT   54
CP/M   98
CPU   6–9, 66, 71, 99, 153, 176
CRC   142
CRT   12, 99, 102–105
CSMA   143

DATA   25–28, 86
data bus   7, 8, 176–177
data logging   166
data tablet   105
database   2, 110, 144–151, 167
datafile   93
debugging   44

decimal (denary) 4
DEF FN 60–61, 62
DELETE 28, 33
DENDRAL 167
device driver 98–99
dialect 17, 25
DIALOG 147
DIALTECH 148
DIM 41, 78
direct mode command 28
discriminant analysis 158–160
disk emulator 89, 92–93
disk formatting 90
disk operating system 11, 89–90,
    97–101
disk organisation 89–90
distribution transformation 68–71
DMA 177
DNA 84, 149–151, 167
double density disk 91
double sided disk 91
DVST 103–105
dynamic dimensioning 41

EBCDIC 9
editor 99, 161
EMBL 149
END 31, 33
ENTER see RETURN KEY
EPB 147
EROS 110
Ethernet 143–144
EXP 52, 62
expert system 151, 167
exponential notation 127

field see record
file commands 96–97
file name 94
file searching 95
file structure 93, 94–96
file type 93, 94–96, 97, 99
flag 72, 170
floating point notation 157, 171–172
floppy disk 11, 90–91
FOR ... NEXT 35–50, 58–59
format descriptor (FORTRAN) 157–158
FORTH 18, 19, 166
FORTRAN 18, 43, 66, 93, 109, 154,
    157–158, 161
function error 172
functions 52–62

GenBANK 149
GET see INKEY$
Gigabyte 92
GIGO 165–166

GINO-F 109
global variable 119
GOSUB 64, 65–67, 75, 119, 162
GOTO 64–65, 72, 75, 162
graph plotting 112–114
graphics commands 109
graphics hardware 102–106
graphics programming 106–109
grey scale 110, 111

hard disk 11, 91–92
hard sectored disk 90
hardware 3, 5
Hardy-Weinberg 131–134
hashing algorithm 96, 100
hexadecimal 5
high level language 16
HOME 43
human error 173

IEEE-488 interface 140
IF ... THEN ... (ELSE) 71–73, 76, 116,
    122
image processing 102, 110–112
IN 178
index variable 36–38
Information Technology 140–151
INKEY$ 81, 84, 86, 87
INPUT 25–26, 32
input device 6, 9–10, 105–106
instruction register 9
instruction set (CPU) 14
INT 52, 54, 61
interactive 22
interface 10, 140, 166, 176–179
interpreter 17, 161
interrupt 177
iteration 127

joystick 106, 178

LEDA 110
LEFT$ 79–80, 87
LEN 79, 80, 86, 87
LET 25, 32
LIFO 19
light pen 106
line number 21, 23, 162, 163
LISP 19
LIST 21
Local Area Network 142–144
local variable 119
LOG 52, 54–55, 57–58, 62
logarithm 54–55
logic error 21
logical device 99
logical operator 73–74, 75, 148

loop 35–51
LSB 171

machine language 14–16, 161
mainframe 6, 152–160
mathematical model 124, 132
matrix arithmetic 107–109
Mbyte 8, 91
Medline 147
memory address 8, 177
memory mapped display 106
microcomputer 6
microdrive 93
microprocessor 7
MID$ 80–81, 87
minicomputer 6
MINITAB 154–156
modem 142, 146
molecular model 112
monitor see operating system
mouse 105
MSB 171
multiplatter drive 92
multitasking operating system 99–100
multiuser operating system 99–100
MYCIN 167

NBRF 149
nested loop 45–50
NEWLINE see RETURN KEY
nibble 5
NOT 73
numeric array 40–45
numeric variables 24–26, 32, 163

octal 5
ON ... GOSUB 67, 68–71, 75
ON ... GOTO 67–68, 75
Online Information Retrieval 146–149
opcode 14–16
OPEN 96
operating system 13, 97–101
optimisation 161–164
OR 73
ORACLE 146
OUT 178
output device 6, 12

parallel transmission 7–8, 140, 176
parity 141
Pascal 17–18, 21, 43, 66, 93, 109, 115,
   139, 161
password 148, 152
physical device 99
PIA 10, 12
pixel 104, 106
pointer see file structure

polar coordinate 106–108
precedence 29, 74–75
PRESTEL 146, 165
PRINT 23, 32, 39–40
print formatting 82–83
PRINT USING 82–83, 87
printer 12
procedure 119
program 8, 13
program command 21
program file 93
programming language 13, 14
pseudocode 118, 120

radian 53
RAM 10–11
RAM disk 92
random access file 97
random number generator 56
RANDOM(IZE) 56
raster scan display 103–105
RCF 22
READ 25, 27–28, 33, 86
record 93–96, 144–145
recording density 91
recursion 66
register (CPU) 7–9
relational operator 71, 75
REM 23, 32
remote sensing 110–111
REPEAT ... UNTIL 116–117, 121, 163
resolution 103–104, 110, 111
RETURN KEY 9, 21–22, 84
RETURN statement 65–66, 71, 75
reverse polish notation 18–19
RIGHT$ 79–80, 87
RND 52, 55–56, 62, 135, 136
ROM 10, 11, 97
rounding error 172
RS-232 interface 140
RUN 21, 24, 26

SECANT 54
sector 89–90, 91
security 152
selection 131–134
semicolon 39, 44, 50
sequential file 97
serial transmission 7, 140
SGN 52, 61
significant digit 44
simulation 124, 132
SIN 52, 54, 61
single density disk 91
soft sectored disk 90
software 3
sorting 82, 119–123

SPSS(–X)   154, 156–160
SQR   52, 62
stack   19
start bit   141
statistics   42–43, 47–50, 66–67, 69–71,
     154, 166
status register   170
Stefan-Boltzmann   125
STEP   37, 38, 58–59
stochastic model   135
stop bit   141
storage device   6, 10–11, 90
STR$ 79, 87
string array   78
string comparisons   81–82
string variables   77–78, 86
stroke/refresh tube   103–104
structured programming   20, 115–123,
     124
subroutines   see GOSUB
subscript   40–41
synchronous transmission   141
syntax error   21
system command   21

TAN   52, 54, 61
tape   11
teletext   146
temporary file   93
textfile   93, 99, 161

token   143
Toxline   147
transformations   107–109
TRACE   48
truth table   74
TTL   176–177
two dimensional array   45–50
two's complement   169–171
type mismatch   78

unary operator   75
unconditional command   64–67
UNIX   100–101, 150
user defined function   see DEF FN

V24 interface   140
VAL   79, 87
VDU   12, 23
vector plotting   103, 104–105
video digitiser   110
Viewdata   146

WHILE ... WEND   116–117, 120, 127,
     163
Winchester disk   92
word   4, 178
word processing   2–3, 165

XOR   73

ZR Online   147